AF598374

Experimentation for Students of Engineering

Volume III

Fluid Mechanics, Thermodynamics, and Heat Transfer

Experimentation for Students of Engineering

Volume I

Experimental Method and Measurement

Volume II

Applied Mechanics

Volume III

Fluid Mechanics, Thermodynamics, and Heat Transfer

EXPERIMENTATION FOR STUDENTS OF ENGINEERING

Edited by

L. Maunder,
Ph.D., Sc.D., C.Eng., F.I.Mech.E.
Professor of Mechanical Engineering, University of Newcastle upon Tyne

S. D. Probert,
M.A., D.Phil., C.Eng., M.I.Mech.E., A.F.R.Ae.S., M.Inst.P., M.Inst.F.
Professor of Engineering Thermodynamics, School of Mechanical Engineering, Cranfield Institute of Technology

Volume III
Fluid Mechanics, Thermodynamics, and Heat Transfer

S. M. Fraser, B.Sc., Ph.D., C.Eng., A.F.R.Ae.S.
Senior Lecturer in the Mechanics of Fluids, University of Strathclyde

R. S. Hill, M.A., D.Phil., C.Eng.
Lecturer in Mechanical Engineering, University of Newcastle upon Tyne

J. F. T. MacLaren, B.Sc., Ph.D., C.Eng., F.I.Mech.E.
Reader in Thermodynamics and Heat Transfer, University of Strathclyde

S. D. Probert

Heinemann Educational Books
London

Heinemann Educational Books Ltd
LONDON EDINBURGH MELBOURNE TORONTO
AUCKLAND JOHANNESBURG SINGAPORE KUALA LUMPUR
HONG KONG IBADAN NAIROBI NEW DELHI

ISBN 0 435 71534 8 (cased)
ISBN 0 435 71535 6 (limp)

First published 1972

Published by Heinemann Educational Books Ltd
48 Charles Street, London W1X 8AH

Text set in 10/11pt. Monotype Times New Roman printed by letterpress, and bound in Great Britain at The Pitman Press, Bath

Editors' Foreword

Experimentation is essential in engineering. But despite the growth of elaborate undergraduate laboratories, experimental tuition has rarely kept pace with the increasing theoretical content of engineering courses. This series of books grew from a belief that the process of learning in engineering laboratories would be enhanced by the ready availability of comment and advice such as is provided in these texts. Although intended primarily for use in universities and colleges of technology, certain sections may also serve as more general references to experimental practice. SI units have been adopted throughout, supplemented in some instances by the British system of units.

Volume I contains an introduction to the general requirements of experimental method, and describes the principles and applications of some modern instruments. Volume II (Applied Mechanics) and Volume III (Fluid Mechanics, Thermodynamics and Heat Transfer) outline sets of experiments which range from the simple to the difficult, and from the first to the final year of an undergraduate course. The outlines are not, of course, intended to be followed slavishly but to offer material on which coherent laboratory courses can be based.

We are indebted to many sources of information, which are recorded in the successive volumes. We record also our appreciation of the patience shown by the publisher in what soon became a complex task. But most of all our thanks are due to the authors for the way in which they have co-operated to realize the original idea. It is they who have written the series and any virtues it may possess are due to them.

1972 L.M.
S.D.P.

Preface to Volume III

Each of the three parts of this volume is presented by the authors in ways in which they believe will be most easily understood by the student. The aim throughout has been to emphasize the basic principles of engineering, illustrating where possible the desirability of having adequate experimental bases for all theoretical analyses. However an experiment should not be attempted until proper planning and an analysis of the available information have been undertaken, so that only the minimum number of time-consuming experimental tests need be made. The limitations of this approach are considered: in Experiment 12, for example, it is shown that subsonic flow theory will not describe 'choking'.

This volume emphasizes the essential unity of fluid mechanics and thermodynamics, which unfortunately are sometimes studied in isolation from one another. The final Experiment of Part II (Thermodynamics) considers thermal radiation, so leading naturally into Part III (Heat Transfer), the last experiment of which provides an introduction to vacuum technology dealing specifically with heat transfers across vacua. The rarefied gas flows involved are usually of such low Reynolds numbers (due to the low densities) that they can be described by laminar flow gas dynamics, which is discussed in Part I (Fluid Mechanics).

For those contemplating the initial year of an undergraduate course, the recommended sequence is Experiments 1 to 5 and 13 to 16 inclusively. In the second year of the course, Experiments 6 to 9, 17 to 20 and 24 to 26 are likely to supplement satisfactorily the theoretical knowledge provided in lectures at that level. For the final year Experiments 10 to 12, 21 to 23, 27 and 28 could be developed into open-ended projects. However the choice is arbitrary and each of the three parts of the text can be used independently.

1972

S.M.F.
R.S.H.
J.F.T.M.
S.D.P.

Contents

Part III Heat Transfer

Experiments 24 to 27 by J. F. T. MacLaren
Experiment 28 by S. D. Probert

Part I

Fluid Mechanics

Introduction

Fluid mechanics deals analytically and experimentally with fluids in motion; the fluid will, in general, be moving in proximity to a solid boundary and in particular the fluid may be contained inside a solid boundary, e.g. pipe flow, or the fluid may be passing a solid body in the field of flow, e.g. flow over an aerofoil.

For internal flows it is necessary to understand the dominant characteristics which govern the flows through the duct, e.g. boundary layer friction, changes in section or direction, etc., and in particular it is required to gauge accurately the rate of flow. The energy expended in moving the fluid can be provided by allowing the fluid to pass from one level to a lower level but more generally a pump is employed as the energy source. Thus not only is it necessary to estimate the energy loss in the duct to decide the size of pump required, but the design of the pump and its characteristics must be understood in order to provide the energy as efficiently as possible. Conversely, where the fluid is used as a source of energy, the design of turbines to abstract energy efficiently must be considered.

For external flows the fluid is usually stationary and a solid body is designed to move through the fluid with as little disturbance as possible, e.g. an aerofoil is designed to give a high lift/drag ratio and in this case the fluid characteristics which dominate the flow are boundary layer growth, transition from laminar to turbulent flow and boundary layer separation. For ships, etc., there is another characteristic to be considered, namely wave generation. In order to move a solid body through a fluid, a thrust must be provided to overcome the drag experienced by the body; this thrust is obtained by doing work on the fluid, so causing a change in momentum of the fluid e.g. by means of propeller or jet.

Fluid flow is designated 'incompressible' if the flow velocity is low in comparison with the local speed of sound (i.e. less than about 40 per cent) and 'compressible' above this figure. For compressible flow the variation of density is significant and hence the flow analysis must be more comprehensive than that for incompressible flow.

1

Viscosity of Fluids

For any system in which a real fluid is moving relative to a solid body, the rate at which movement occurs is a direct function of the dynamic viscosity, μ, of the fluid. For any fluid in relative motion, shear stresses opposing the motion are set up in the fluid, the ratio of shear stress to rate of shear strain being defined as the viscosity of the fluid.

For homogeneous fluids the viscosity is a simple function of temperature and virtually independent of pressure. For a liquid, viscosity decreases with increasing temperature whereas for a gas it increases with temperature.

Redwood Viscometer

Theory

There are two main types of commercially available viscometer, one of which gives a comparative measurement relative to some standard fluid while the other provides an absolute measurement of viscosity. With the former the period required for a determined volume of liquid to pass through a short length of narrow bore tube, which has accurately known dimensions, is measured. In such an instrument this period is related to the viscosity in the following manner

$$\nu \equiv \mu/\rho = AP - B/P$$

where ν is the kinematic viscosity ($m^2\ s^{-1}$),
μ is the dynamic viscosity ($kg\ m^{-1}\ s^{-1}$),
ρ is the density ($kg\ m^{-3}$),
P is the period for passage of given volume (s),
and A and B are instrumental constants found by calibrating with a standard fluid.

Calibrations at the National Physical Laboratory have given the following values of A and B for the Redwood number 1 viscometer:

$40 \leqslant P \leqslant 85$ s $\quad A = 0{\cdot}00264, \quad B = 1{\cdot}90$
$85 \leqslant P \leqslant 2000$ s $\quad A = 0{\cdot}00247, \quad B = 0{\cdot}65$

The Redwood viscometer is widely used because of its simplicity of operation. The relative viscosity obtained from a test is quoted in 'Redwood Seconds', i.e. the time taken for a measured volume to pass through the viscometer.

Experimental Details

Test Equipment. The Redwood viscometer consists of an inner cylinder into which the liquid to be tested is poured and an outer cylinder which usually contains water to maintain the liquid at a specified temperature throughout the test. A hook gauge is attached to the inner cylinder so that it can be

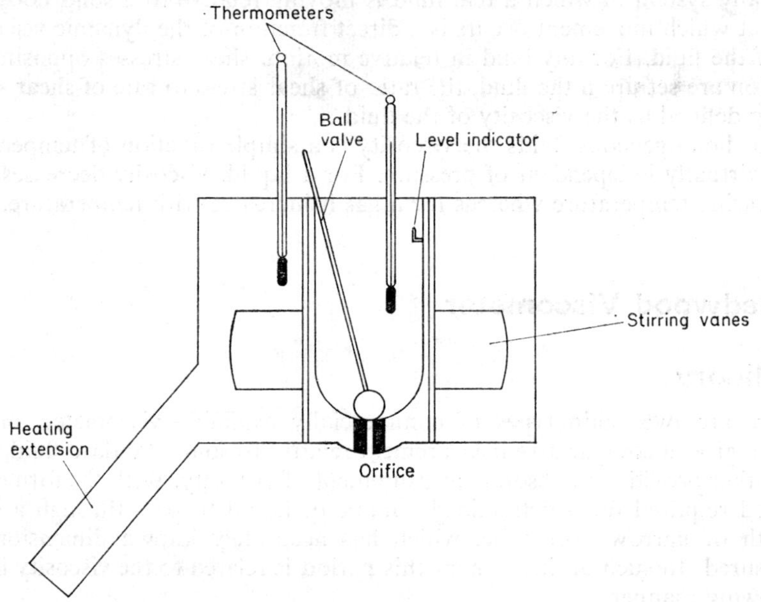

FIGURE 1.1 Redwood viscometer.

filled to the same depth of fluid for each test. A narrow-bore orifice is drilled in the agate seat which is cemented in the base of the inner cylinder, a silver-plated ball valve being used to start and stop the flow. A cylinder with vanes for stirring the water is fitted on the inner cylinder which also serves to support the bath thermometer. The water bath has a side projection to enable small temperature increases to be made by direct heating.

Procedure. Heat sufficient water to approximately 3 °C in excess of the test temperature and pour it into the water jacket. Heat the test liquid to the required temperature and pour it carefully into the inner cylinder until the hook gauge is just covered. When the temperature of both liquids is the same, note the value of the temperature and measure the period required to discharge 50 ml.

Repeat the test throughout the required range of temperature. On each occasion calculate the viscosity of the test liquid from the instrument constants and the viscosity and specific gravity of the standard liquid. Plot a graph of viscosity *versus* temperature.

Discussion

Comment upon the errors involved with reference to the volume of test liquid used, timing of the discharge, variation of the relative density of the test liquid with temperature and the effect of impurities, and changing the orifice dimensions.

Searle's Viscometer

Theory

Direct measurement of viscosity is usually obtained by torque measurement on a fluid in shear or by a more complicated version of the capillary viscometer. Considering the torque measurement method it can be shown that an immersed body, e.g. a cylinder, when rotated at constant angular velocity experiences a viscous retardation on its surface; this retardation can be equated to the torque about the axis of symmetry of the body, which is required to maintain a steady state.

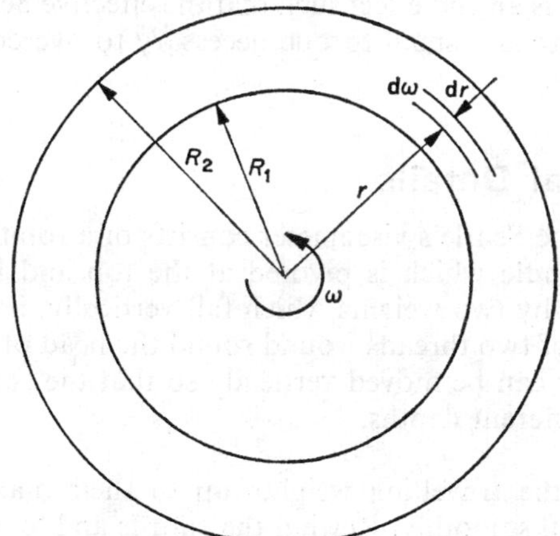

FIGURE 1.2 Annulus of liquid in couette flow.

Consider two concentric cylinders, height h, as shown in Figure 1.2, the outer fixed and the inner rotating with constant angular velocity ω. The condition of zero slip at the solid boundaries means that the liquid filling the gap between the two cylinders will have zero angular velocity at radius R_2 and ω at radius R_1.

Consider an elementary ring of radius r, width dr, and let the change in angular velocity be dω across the ring. Thus

$$\text{viscous stress } \tau = \mu \frac{dv}{dr}$$

$$= \mu r \frac{d\omega}{dr}$$

$$\text{viscous torque} = \tau \times \text{area} \times \text{radius}$$

$$= \mu r \frac{d\omega}{dr} 2\pi rhr$$

$$= \mu 2\pi hr^3 \frac{d\omega}{dr}$$

For the steady state the applied torque T must exactly balance the net viscous torque:

$$\int_0^\omega d\omega = -\frac{T}{\mu \,.\, 2\pi h}\int_{R_2}^{R_1}\frac{dr}{r^3}$$

$$\mu = \frac{T}{4\pi\omega h}\left[\frac{1}{R_1^2} - \frac{1}{R_2^2}\right] \tag{1.1}$$

Hence μ can be calculated from the measured torque and angular velocity and the known dimensions of the cylinders.

In practice there is an end effect such that the effective height of the cylinder is $(h + l)$, and there is a small torque necessary to overcome the friction in the pulley bearings.

Experimental Details

Test Equipment. The Searle's viscometer consists of a rotatable inner cylinder mounted on a spindle which is pivoted at the top and bottom. The inner cylinder is rotated by two weights which fall vertically, imparting rotational motion by means of two threads wound round the head of the inner cylinder. The outer cylinder can be moved vertically so that the rotating cylinder can be immersed to different depths.

Procedure. Wind the travelling weights up to their maximum height and ensure that they fall smoothly. Rewind the chords and lock them in position. Fill the outer cylinder with the test liquid and position it to give the maximum depth of immersion of the inner cylinder. Unlock the inner cylinder and, when the motion is steady, note the time taken for a given number of revolutions. Repeat with two different weights attached.

Repeat the above procedure for several depths of immersion.

Tabulate values of the weight, W, time per revolution, t, depth of immersion, h, and product Wt.

Plot a graph of l/t against W and hence estimate the torque required to overcome friction.

Plot a graph of h against actual torque T and hence estimate a value for l, the end correction factor.

Calculate the absolute viscosity of the test liquid.

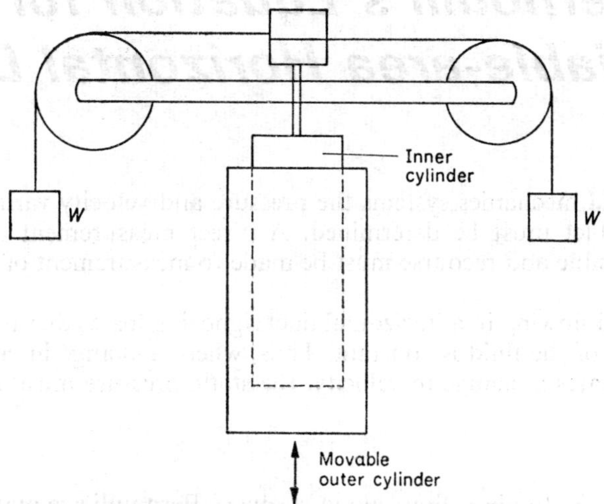

FIGURE 1.3 Principle of Searle's viscometer.

Discussion

Explain how estimates of the torque required to overcome bearing friction and the end correction factor are obtained from the graphs.

Discuss the difficulties that would be encountered in testing low viscosity liquids by this method.

Indicate the type and probable magnitude of errors in this test.

2

Bernoulli's Equation for a Variable-area Horizontal Duct

In many fluid mechanics systems the pressure and velocity variation along a horizontal duct must be determined. A direct measurement of velocity is often impossible and recourse must be made to measurement of pressure and temperature.

For a fluid flowing in a horizontal duct, ignoring losses due to friction, the total energy of the fluid is constant. Thus, where a change in cross-sectional area necessitates a change in velocity, the static pressure must also change.

Theory

Considering frictionless flow along a duct, Bernoulli's equation for any streamline may be written as

$$\frac{p}{\rho} + \frac{v^2}{2} + gZ = \text{constant} \tag{2.1}$$

where p is the static pressure, ρ the density, v the velocity, g the acceleration due to gravity and Z the height above the datum. The value of the constant may vary from streamline to streamline.

For a horizontal duct, the Z term is constant and we may therefore write

$$\frac{p}{\rho} + \frac{v^2}{2} = \text{constant} \tag{2.2}$$

and since the ρ is constant for an incompressible fluid it follows that

$$p + \tfrac{1}{2}\rho v^2 = \text{constant} \tag{2.3}$$

Equation (2.3) is a form of Bernoulli's equation which is often used in aerodynamics. The constant is referred to as the total pressure, p_0, and the term $\frac{1}{2}\rho v^2$ as the dynamic pressure, i.e.

static pressure + dynamic pressure = total pressure.

The total pressure is the value of the static pressure when the fluid is brought to rest without losses. Thus if we can have a system which measures the static and the total pressure of a moving fluid we can find the dynamic pressure and hence the velocity:

$$v = \left[\frac{2(p_0 - p)}{\rho}\right]^{\frac{1}{2}} \tag{2.4}$$

where v is in m s^{-1}, p in N m^{-2}, and ρ in kg m^{-3}.

This can be achieved with an instrument called a pitot-static tube (Figure 2.1). When the tube is aligned in the direction of flow, the fluid is brought

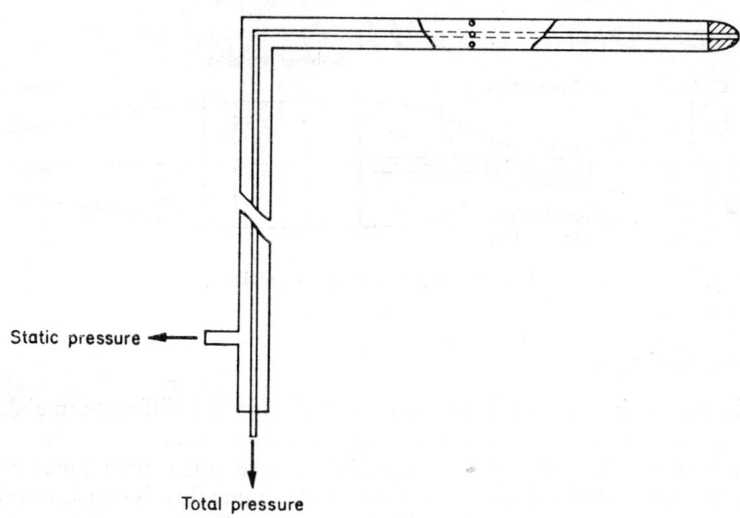

FIGURE 2.1 Pitot-static tube.

to rest in the inner tube so that the total pressure can be measured. The outer tube enables the static pressure to be obtained. In general the instrument will measure total pressure accurately over a range of yaw angles (i.e., inclinations to the main flow direction) but the static pressure holes must be very critically placed relative to the nose of the instrument, which must be carefully aligned in the flow.

Experimental Details

Test Equipment. The general layout of the test equipment is shown in Figure 2.2. Air is passed into the reservoir from a fan and a coarse and fine wire-mesh screen aligns the flow before it passes through the Perspex contraction section. A detachable diffuser section can be positioned at the exit. The pitot-static tube is clamped to a traversing slide and is set up to measure the pressures at the centre line streamline.

Procedure. Connect the pitot-static tube to manometers so that static, dynamic and total pressure can be read. Remove the diffuser section and set the control valve to give maximum flow. Take the manometer readings for six positions in the contraction and note the depth of the section at the static hole position. Repeat for reduced flow, and then repeat the whole procedure with the diffuser section attached. If necessary reconnect the static pressure connection to read pressures below atmospheric for the diffuser test.

Plot graphs of static, dynamic and total pressure along the contraction for the four conditions.

Plot graphs of (i) velocity, and (ii) cross-sectional area times velocity along the contraction.

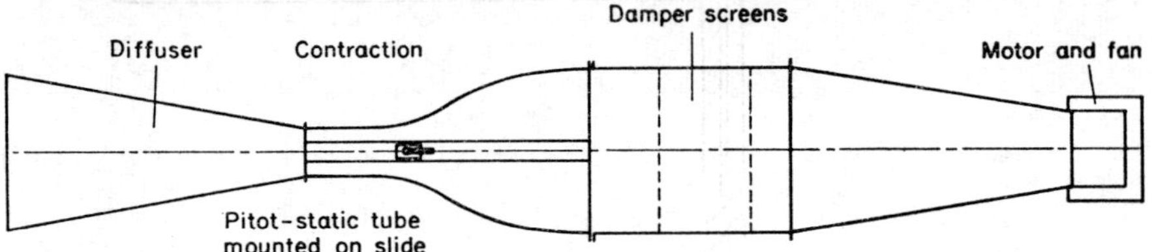

FIGURE 2.2 Variable-area duct.

Discussion

Why is the static pressure below atmospheric for the diffuser-attached case and why should this increase the velocity and flow rate?

Comment on the shape of the curve of cross-sectional area times velocity with reference to the condition that the actual volume flow is constant through the duct for each fan setting. Explain the apparent discrepancy in terms of the likely velocity distribution across the duct.

Make an assessment of the possible errors in this experiment in terms of pitot-static tube alignment and manometer readings.

3

Bernoulli's Equation Applied to a Vertically Displaced Pipe Containing Water

In systems with steady flow through pipes of constant cross-sectional area the kinetic energy, based on mean flow velocity, is constant at all points. Assuming that the loss of pressure due to friction may be neglected in comparison with the other changes, then vertical changes in pipe position will cause a change in potential energy which is exactly balanced by the change in the fluid static pressure (Figure 3.1).

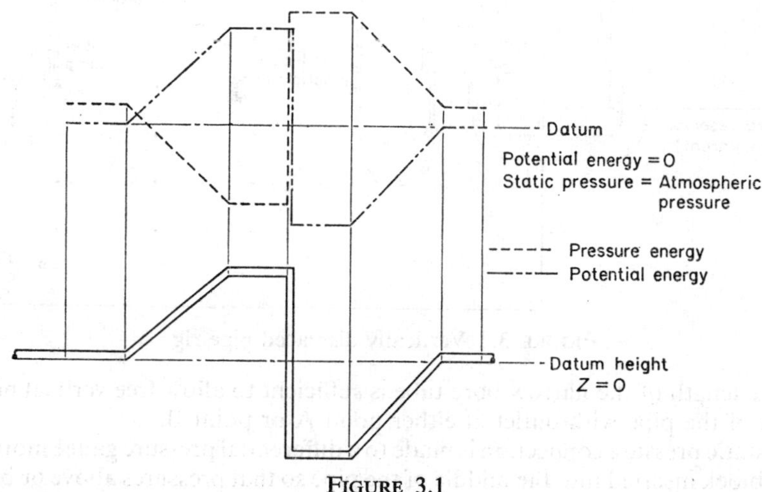

FIGURE 3.1

Theory

For a constant-diameter pipe $v^2/2$ is constant so that Bernoulli's equation now becomes

$$\frac{p}{\rho} + gZ = \text{constant} \tag{3.1}$$

In the absence of losses due to viscosity, if the potential head Z is increased by raising the pipe, the static pressure head will fall. Conversely, if the potential head is decreased the static pressure head will increase.

When a pipe containing a liquid has a portion of its length with a fluid static pressure lower than atmospheric the system is referred to as a syphon. Most liquids contain dissolved gases, and the reduction in pressure occurring in a syphon may be sufficient to cause the gases to come out of solution or for vapour pockets to form; this is referred to as cavitation.

Experimental Details

Test Equipment. A sealed reservoir tank (Figure 3.2) feeds a constant level header tank from which a narrow bore tube conveys the water to an outlet some distance below the header tank level.

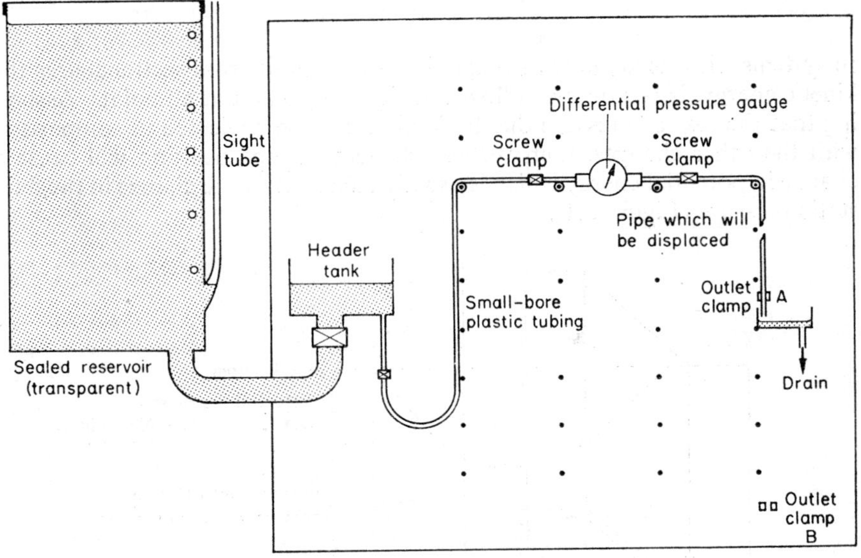

FIGURE 3.2 Vertically displaced pipe rig.

The length of the narrow bore tube is sufficient to allow free vertical movement of the pipe with outlet at either point A or point B.

A static pressure connection is made to a differential pressure gauge mounted on a block inserted into the middle of the pipe so that pressures above or below atmospheric can be measured.

Screw clamps are fixed before and after the pressure tapping to alter the local pressure and overall flow rate.

Procedure. Ensure that the main reservoir is properly sealed and slowly open the tap leading to the header tank; observe the way in which the water level in the sight tube moves as the header tank is filled.

Clamp the pipe exit at position A, place the pipe on the lowest set of wooden plugs and arrange the pressure tapping to read a pressure in excess of atmospheric. Open the control valve at the header tank and, when the flow is steady, note the pressure reading and the distance between the pipe centre-line and the water level in the header tank.

Repeat the above procedure for three readings below the water level and three readings above the water level, noting that the pressure tapping must be arranged to read sub-atmospheric pressure when the pipe is above the water level in the header tank.

Clamp the outlet at position B and, with the pipe in the top set of pegs, attach a screw clamp to the pipe before the pressure tapping. Take a reading with maximum flow and with the clamp screwed in to give approximately the same flow rate as in the previous test. Repeat this procedure with the clamp positioned after the pressure tapping.

Plot a graph of pressure head *versus* pipe position relative to header-tank water level. Tabulate the values for the second test.

Discussion

Explain on the basis of simple manometry why the sealed reservoir allows the water level in the header tank to remain constant.

Discuss the interchange between potential head and pressure head.

Comment on the effect of changing the outlet from A to B and on the effect of clamping before and after the pressure tapping point.

Discuss the source and magnitude of possible errors.

4

Thrust-Momentum Relationship for an Air Jet

For many flow problems it is necessary to use Newton's second law of motion in addition to the continuity and energy equations in order to obtain a solution. The method can be applied to find the force acting on a bend say, or, for a free jet, to find the thrust developed.

For a simple system, such as a ducted fan taking air in at atmospheric pressure and exhausting to atmospheric pressure, the net thrust developed is directly equal to the rate of change of fluid momentum.

Theory

Newton's second law states that the net force F in a given direction is proportional to the rate of change of momentum, i.e.

$$\vec{F} = \frac{\mathrm{d}}{\mathrm{d}t}(m\vec{v}) \tag{4.1}$$

where force and velocity v are vector quantities.

For steady flow the mass flow m is constant and thus

$$\begin{aligned}\vec{F} &= \text{mass flow} \times \overrightarrow{\text{change in velocity}} \\ &= m\,\mathrm{d}\vec{v} \\ &= \rho Q\,\mathrm{d}\vec{v}\end{aligned}$$

where F is in newtons, density ρ in kg m^{-3} and volume flow Q in m^3 s^{-1}.

The net force includes pressure, viscous, and body forces acting on the fluid. In this simple case the net force is the thrust developed neglecting viscous effects.

Since the fluid velocity at a point is a function of the pipe radius, the rate of change of momentum will also be a function of radius.

Consider an elemental ring of width dr at radius r where the local velocity is v. Then for the element

$$\text{volume flow } \mathrm{d}Q = v \,.\, 2\pi r \,\mathrm{d}r$$

$$\text{momentum flux } \mathrm{d}M = \rho v^2 \,.\, 2\pi r \,\mathrm{d}r$$

$$\text{total volume flow} = \int_0^R 2\pi v r \,\mathrm{d}r \qquad (4.2)$$

$$\text{total momentum flux} = \int_0^R 2\pi \rho v^2 r \,\mathrm{d}r \qquad (4.3)$$

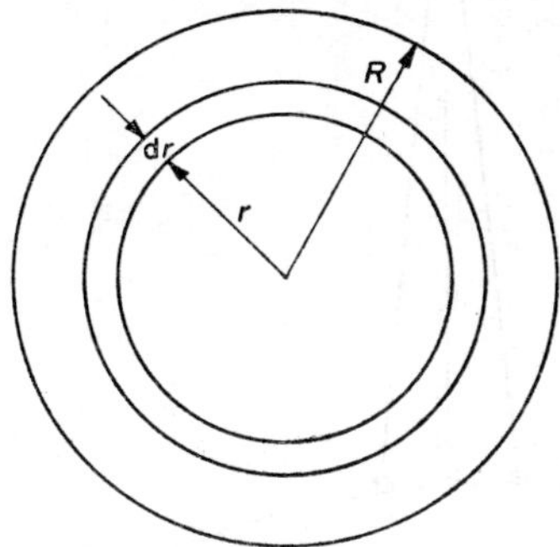

FIGURE 4.1

Experimental Details

Test Equipment. The ducted fan (Figure 4.2) is driven by a variable-speed electric motor and is suspended by two wires from a pivot. An indicating pointer is attached to the casing and moves over a graduated scale which is calibrated in degrees.

The jet of air is traversed by means of a pitot-static tube and the ducted fan can be locked in position whilst the traversing procedure is being completed.

The total weight W of the ducted fan and suspension is required.

Procedure. With the ducted fan in the locked position, run up the motor to full speed and make a complete velocity traverse through the jet using a pitot-static tube at a position where the static pressure through the jet is at atmospheric pressure. (Check by using the static tapping on the pitot-static tube). Unlock the ducted fan and note the angle θ it assumes under the action of the thrust.

Repeat the above procedure for two other speeds.

Tabulate values of θ, $W \sin \theta$, measured thrust and approximate thrust, where

$$\text{measured thrust} = \int_0^R 2\pi \rho v^2 r \,\mathrm{d}r$$

$$\text{approximate thrust} = \rho Q \bar{v}$$

$$\text{mean velocity at exit } \bar{v} = Q/A$$

and
$$Q = \int_0^R 2\pi v r \, dr$$

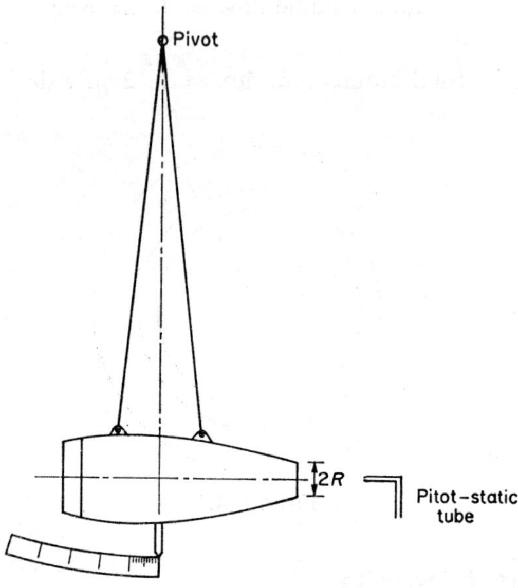

FIGURE 4.2 Jet reaction.

Discussion

Explain the symmetry of the velocity profile and the reason for choosing a station where the static pressure had returned to atmospheric pressure. Hence modify the momentum equation to take account of the effect of static pressure variation across the jet.

Compare the values of $W \sin \theta$, measured thrust and approximate thrust, and state why the approximate thrust should be different from the measured thrust.

Discuss the errors encountered in this test.

5

Flow Measurement by Venturi Meter and Orifice Plate

The measurement of the rate of flow of a fluid is a primary requirement for many systems, and an instrument capable of measuring a pressure differential is the basis of most flow meters. As has been shown in Experiment 2, a change in static pressure can be related to the change in velocity caused by a variation of the cross-sectional area of the flow. Thus, locally constricting the flow causes a measurable change in pressure.

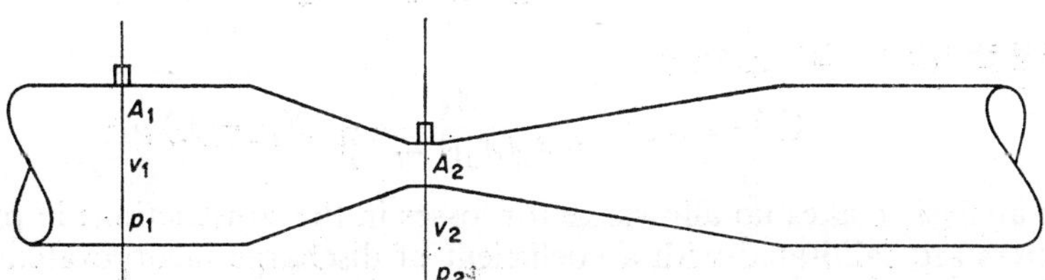

FIGURE 5.1 Venturi meter

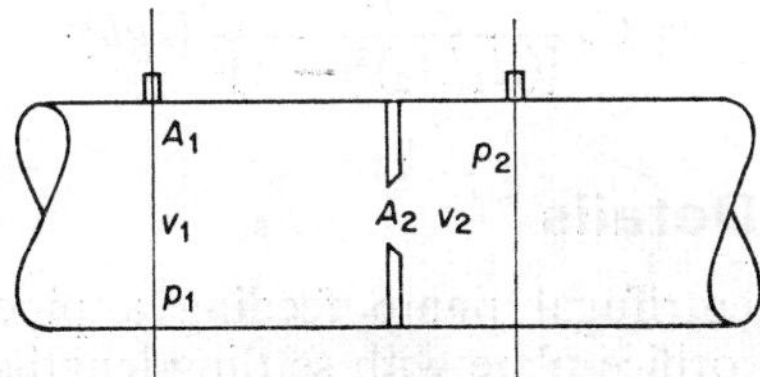

FIGURE 5.2 Orifice plate

Theory

Consider the Venturi meter and orifice plate shown in Figures 5.1 and 5.2. The continuity equation is

$$A_1 v_1 = A_2 v_2$$

where A_1, A_2 are the cross-sectional areas, v_1, v_2 the velocities at the respective cross sections. Hence

$$v_2 = \frac{A_1}{A_2} v_1 \tag{5.1}$$

The energy equation for the horizontal meter is

$$\frac{p_1}{\rho} + \frac{v_1^2}{2} = \frac{p_2}{\rho} + \frac{v_2^2}{2}$$

where p is pressure.

$$\frac{p_1 - p_2}{\rho} = \frac{v_2^2 - v_1^2}{2}$$

$$= \frac{v_1^2}{2}\left[\left(\frac{A_1}{A_2}\right)^2 - 1\right]$$

$$\frac{v_1^2}{2} = \frac{(p_1 - p_2)/\rho}{(A_1/A_2)^2 - 1}$$

$$v_1 = \left[\frac{2(p_1 - p_2)/\rho}{(A_1/A_2)^2 - 1}\right]^{\frac{1}{2}} \tag{5.2}$$

where $h = (p_1 - p_2)/w$. Thus

$$Q = A_1 v_1 = \frac{A_1}{[(A_1/A_2)^2 - 1]^{\frac{1}{2}}} [2(p_1 - p_2)/\rho]^{\frac{1}{2}} \tag{5.3}$$

The analysis makes no allowance for losses in the constriction: in practice all meters are calibrated with a coefficient of discharge incorporated in the equation as follows:

$$Q_{\text{actual}} = C_d Q_{\text{theoretical}}$$

$$= C_d \frac{A_1}{[(A_1/A_2)^2 - 1]^{\frac{1}{2}}} [2gh]^{\frac{1}{2}} \tag{5.4}$$

Experimental Details

Test Equipment. A centrifugal pump feeding a pipeline incorporating a Venturi meter and an orifice plate with settling lengths before and after the meters (Figure 5.3). The differential pressure across each meter is measured with a manometer and the actual volume flow rate is determined by collecting the water in a measuring tank over an appropriate time interval.

Procedure. Start the pump with the control valve fully closed and then gradually open the valve until maximum flow is obtained. When the flow is steady, take the readings on the differential manometers and measure the flow rate directly. Note the water temperature.

Repeat the procedure for five other flow rates.

Tabulate differential pressure, theoretical discharge, discharge coefficient and Reynolds' number for each meter, and plot a graph of C_d *versus* Reynolds' number.

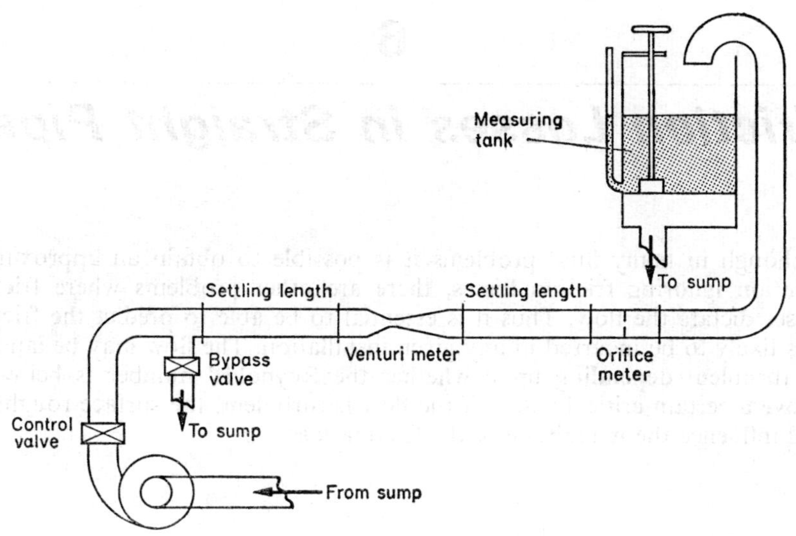

FIGURE 5.3 Flow meters in test rig.

Discussion

Why is the discharge coefficient so different for the two meters? Explain why there should be a Reynolds' number effect on the discharge coefficient and why it is more apparent for the orifice meter. What happens to the discharge coefficient as the area ratio is changed?

Indicate types and magnitudes of likely errors. How would you calibrate a flow meter in an airflow rig?

6

Friction Losses in Straight Pipes

Although in many fluid problems it is possible to obtain an approximate solution ignoring friction losses, there are other problems where friction losses dictate the flow. Thus it is essential to be able to predict the friction loss likely to be incurred in any given installation. The flow may be laminar or turbulent depending upon whether the Reynolds' number is below or above a certain critical value. If the flow is turbulent, the surface roughness will influence the magnitude of the friction loss.

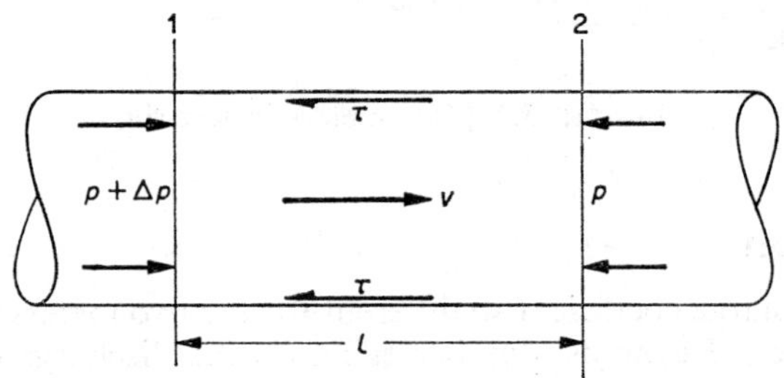

FIGURE 6.1 Pipe flow.

Theory

Consider the flow through a pipe of diameter d, where the mean velocity of flow is v. Between sections 1 and 2 the action of the viscous shear stress will cause the pressure to drop from $p + \Delta p$ to p over the length l.

If the flow is steady the net force must equal the shear force, i.e.

pressure difference × cross-sectional area = shear stress × wetted area

$$\Delta p \,.\, \tfrac{1}{4}\pi d^2 = \tau \,.\, \pi d l$$

$$\Delta p = 4 l \tau / d \tag{6.1}$$

We define a friction factor f,

$$f = \frac{\tau}{\frac{1}{2}\rho v^2} = \frac{\text{shear stress}}{\text{dynamic pressure}} \tag{6.2}$$

Hence

$$\Delta p = \frac{4l}{d} \cdot \frac{\rho v^2 f}{2}$$

$$\Delta p = 4f\frac{l}{d} \cdot \tfrac{1}{2}\rho v^2$$

Thus for a given pipe, the pressure loss due to friction is directly proportional to the dynamic pressure.

It can be shown by dimensional analysis that the friction factor f is a function of Reynolds' number R and surface roughness ratio λ/d, where λ is the mean height of surface irregularities and d is the pipe diameter.

Let the shear stress τ be some function of velocity v, density ρ, viscosity μ, diameter d and surface roughness λ. Hence

$$\tau = \phi(v, \rho, \mu, d, \lambda)$$

where ϕ is some function. Let

$$\tau = kv^a \rho^b \mu^x d^y \lambda^z$$

where k is a constant. Thus

$$\frac{\mathrm{M}}{\mathrm{LT}^2} \equiv \left(\frac{\mathrm{L}}{\mathrm{T}}\right)^a \left(\frac{\mathrm{M}}{\mathrm{L}^3}\right)^b \left(\frac{\mathrm{M}}{\mathrm{LT}}\right)^x (\mathrm{L})^y\, (\mathrm{L})^z$$

Equating indices,

for M $\quad 1 = b + x$

for L $\quad -1 = a - 3b - x + y + z$

for T $\quad -2 = -a - x$

Therefore

$$\begin{aligned} b &= 1 - x \\ a &= 2 - x \\ y &= -1 - a + 3b + x - z \\ &= -x - z \end{aligned}$$

Hence

$$\begin{aligned} \tau &= kv^{2-x}\rho^{1-x}\mu^x d^{-x-z}\lambda^z \\ &= k\rho v^2 \left(\frac{\mu}{\rho v d}\right)^x \left(\frac{\lambda}{d}\right)^z \end{aligned}$$

$$\frac{\tau}{\tfrac{1}{2}\rho v^2} = \phi\left(R, \frac{\lambda}{d}\right) \tag{6.3}$$

Experimental Details

Test Equipment. Four long pipes, Figure 6.2, are connected in parallel between two header tanks. Water is supplied by a centrifugal pump and a

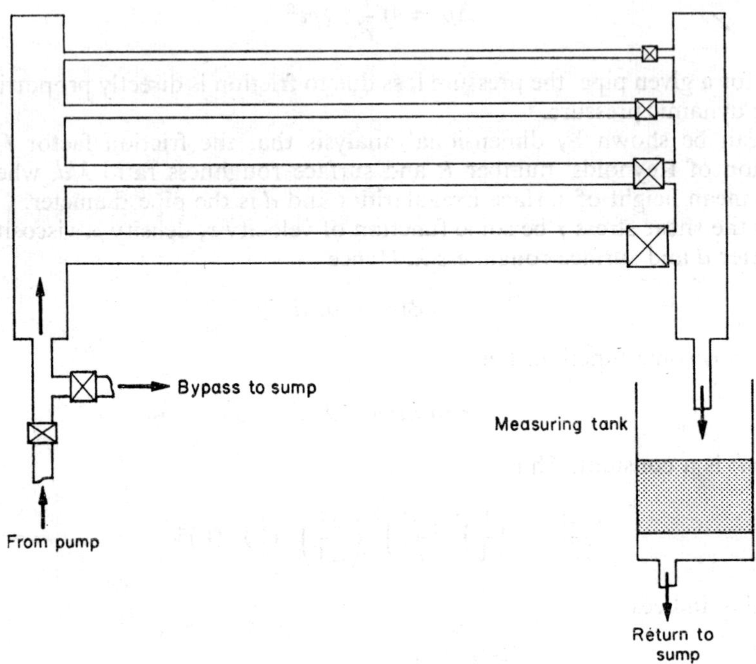

FIGURE 6.2 Friction in pipes.

bypass circuit is incorporated to control the flow through the pipes. Pressure-tapping points are situated at the inlet and outlet to the pipes at a known distance apart; the inlet point is situated sufficiently far away to ensure a fully-developed boundary layer. The outlet header tank discharges into a measuring tank so that the volume flow can be determined directly.

Procedure. Start the pump with the outlet valve fully closed and then gradually open this valve to give maximum flow. Open the inlet valve to one of the pipes and make sure that the other three are closed. When the flow is steady take the pressure reading on the manometer and measure the flow rate. Repeat the procedure for five other flow rates using the main valve and the bypass valve to give the required running conditions. Note the water temperature.

Repeat the procedure for the other three pipes.

Tabulate values of friction pressure loss, friction factor and Reynolds' number.

Plot graphs of friction factor *versus* Reynolds' number and compare them with graphs obtained by plotting $f = 16/R$ for laminar flow and $f = 0{\cdot}079R^{-\frac{1}{4}}$ for turbulent flow.

Discussion

Comment upon the difference between the two types of flow, i.e. laminar and turbulent, on the basis that the Reynolds' number is the ratio of inertia force to viscous force.

Indicate the type and magnitude of possible errors occurring in this test.

Deduce the effect on friction pressure loss of doubling the pipe diameter for a given rate of flow. Use the equation for Δp, given in the theory, and substitute in the friction factor equations for laminar and turbulent flow. Discuss the relevance of this type of analysis to industrial practice.

7

Losses at Sudden Changes in Section and Direction

In the previous experiment the only loss considered was due to frictional shear stress at the wall of the pipe. In general, however, a pipe run will include changes in diameter as well as bends. In this experiment losses associated with the main types of change of section and direction will be investigated, including sudden enlargement, sudden contraction and various shapes of bend.

Theory

Consider the case of a sudden enlargement, as shown in Figure 7.1. The jet of fluid issuing from the smaller pipe will tend, due to its inertia, to continue to flow in the larger pipe with the same cross-sectional area. However, due

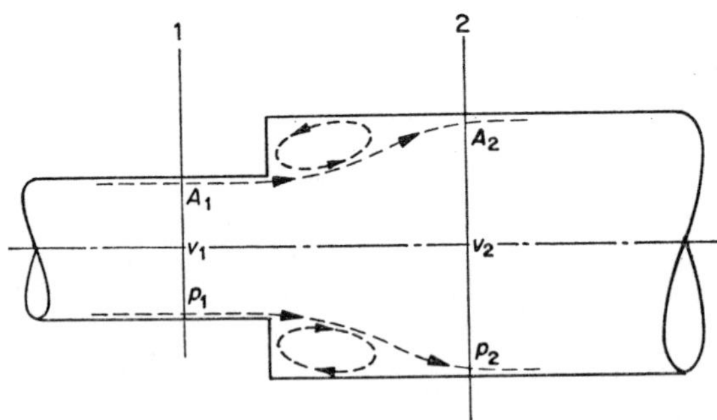

FIGURE 7.1 Discontinuity in pipe diameter.

to viscous action at the jet boundary, fluid is entrained from the second pipe and a recirculating vortex system is set up as shown. The jet spreads out and at section 2 of the pipe it completely fills the larger pipe.

The continuity equation is

$$A_1v_1 = A_2v_2 \tag{7.1}$$

where A_1, A_2, v_1, v_2 are cross-sectional areas and velocities of flow at the respective cross sections. Also

Net force = Rate of change of momentum

The pressure in the vortex region is not precisely equal to p, but we make this assumption in order to simplify the analysis. Thus

$$ {}_1A_1 + p_1(A_2 - A_1) - p_2A_2 = (\rho A_2v_2)(v_2 - v_1)$$

$$p_1 - p_2 = \rho(v_2^2 - v_1v_2)$$

$$\frac{p_1 - p_2}{\rho} = v_2^2 - v_1v_2 \tag{7.2}$$

where ρ is density.

The actual difference in pressure is measured by means of tapping points at sections 1 and 2.

The energy equation for the horizontal duct is

$$\frac{p_1}{\rho} + \frac{v_1^2}{2} = \frac{p_2}{\rho} + \frac{v_2^2}{2} + \text{losses}$$

$$\therefore \text{ losses} = \frac{p_1 - p_2}{\rho} + \frac{v_1^2 - v_2^2}{2}$$

$$= v_2^2 - v_1v_2 + \frac{v_1^2 - v_2^2}{2}$$

$$= \frac{(v_1 - v_2)^2}{2} \tag{7.3}$$

i.e. the loss in total pressure between sections 1 and 2 is given by

$\frac{1}{2}\rho$ (difference in velocity)^2

The sudden contraction can be treated in a similar fashion.

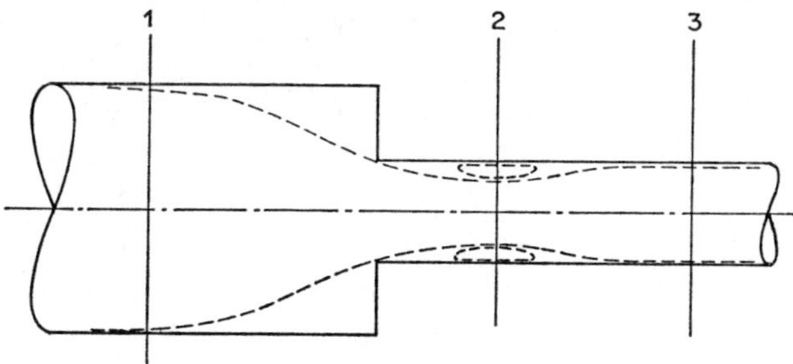

FIGURE 7.2 Sudden contraction in pipe diameter.

In general a fluid will contract with only a small loss and it can be assumed that the major loss occurs between sections 2 and 3.

As the fluid passes from the larger pipe to the smaller pipe it will continue to contract and will form a throat (a *vena contracta*) at section 2. Thereafter, due to viscous entrainment, the jet will expand and fill the pipe at section 3.

The loss is assumed to be given by

$$\text{loss} = \frac{(v_2 - v_3)^2}{2}$$

but, by continuity,

$$A_2 v_2 = A_3 v_3$$

where

$$A_2 = C_c A_3$$

Therefore

$$v_2 = v_3 / C_c$$

and so

$$\text{loss} = \frac{v_3^2}{2}\left[\frac{1}{C_c} - 1\right]^2 \qquad (7.4)$$

The loss incurred at a bend depends on the angle turned through and the shape of the bend. In general it is not possible to derive an analytical expression for bend losses and it is usual to quote empirical values for different shapes and angles in the form of a loss coefficient k where the loss is given by $kv^2/2$.

Experimental Details

Test Equipment. The layout of the test equipment is shown in Figure 7.3. A centrifugal fan drives air through the ducting, which consists of an orifice flow meter, a sudden enlargement, a sudden contraction, a radius bend and a mitre bend with suitable settling lengths in between. An adjustable baffle plate is fitted to the exit so that a range of speeds can be tested.

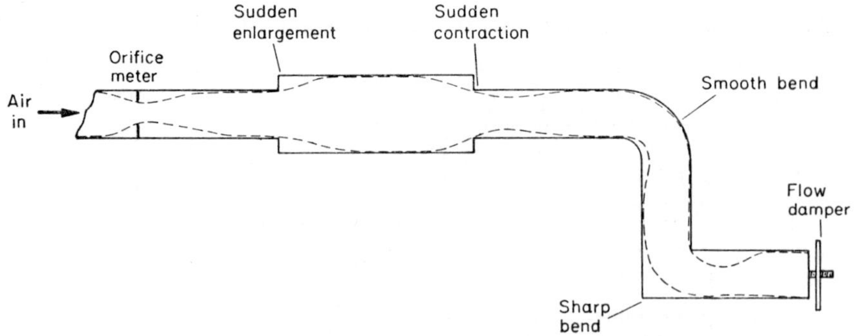

FIGURE 7.3 Sudden changes in section and direction.

Because of the distorted nature of the flow after a bend it is insufficient to measure the change in static pressure at the wall surface, and therefore an impact tube is used with a number of holes drilled in it to measure the average total pressure before and after the bend.

Procedure. Before starting the fan ensure that the manometers are connected to read the correct differential pressure, i.e. an increase in static pressure for the sudden enlargement and a decrease in static pressure for the sudden contraction: the two bends will obviously show a decrease in total pressure due to the bend.

With the exit damper fully open, start the fan and allow the flow to become steady. Note the flowmeter reading and the readings for the sudden changes in section and direction. Note room temperature and pressure. Repeat this above procedure for four other velocities.

From the given value of the flowmeter discharge coefficient evaluate the volume flow for each run and hence evaluate the mean velocity before and after each change in section.

Tabulate velocity, dynamic pressure, theoretical loss in total pressure, change in static pressure, and theoretical change in static pressure for the sudden changes in section. Tabulate also velocity, dynamic pressure, actual loss in total pressure, and loss coefficient k for the bends.

Discussion

Compare the measured results with the theoretical results for the sudden changes in section and discuss the relevance of the assumptions made in the theoretical analysis. Why are the sudden enlargement results likely to show a bigger discrepancy than the sudden contraction data?

Explain how the positioning of the static pressure tapping points could affect the results, and indicate how the readings would be changed by moving the points nearer to or further away from the junction.

Discuss the way in which the number of holes in the impact tubes used for the bends will affect the readings. Is the effect likely to be more pronounced after the bend than before it? Comment on the magnitude of the loss factor and the practicability of using larger radius bends commercially in order to reduce losses.

Discuss the sources and magnitudes of the possible errors.

8

Pressure Distribution Around an Aerofoil

When fluid flows relative to a solid body, either externally, as for an aircraft, or internally, as for pipe flow, the total force experienced by the body is made up of two components, a 'shear' force due to tangential shear stresses and a 'pressure' force due to pressure acting normal to the body surface.

It is possible to assess the shear component in terms of a friction factor, which is a function of Reynolds' number, but if the 'pressure' force is to be found it becomes necessary to consider the effective pressure distribution on the body surface. For flow inside a body, e.g. through a reducing bend, the resultant force and its line of action can be determined from considerations of continuity, energy and momentum. When the flow is external to the body as for an aerofoil, it is not possible to state how much of the fluid is influenced by the body. Hence the pressure distribution is usually measured directly and then integrated to find the resultant force.

Theory

Consider an aerofoil inclined at an angle of incidence α to the free-stream direction (see Figure 8.1).

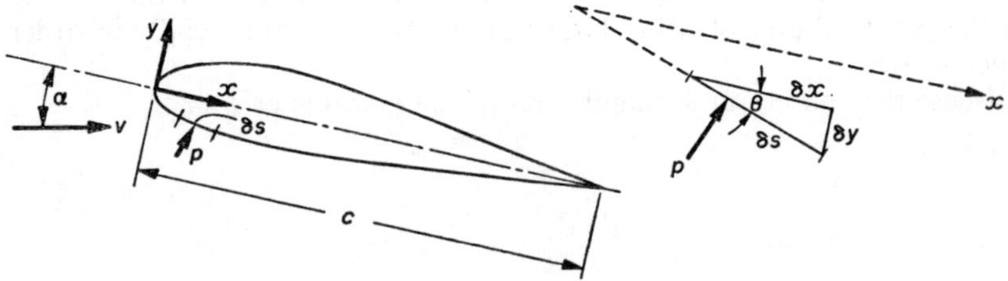

FIGURE 8.1 Aerofoil at incidence α to free-stream.

The chord line of an aerofoil is defined as the line joining the centres of curvature of the leading and trailing edges and has a length c.

The shape of the aerofoil is given in terms of cartesian coordinates x and y which are respectively parallel and perpendicular to the chord line, as indicated.

Consider a small element δs, of the lower surface which has a pressure p acting normal to it, where the pressure is measured relative to the free-stream static pressure and hence may be positive or negative. Assuming unit width, the pressure force acting on the element δs is given by

$$\delta F = p\delta s 1$$

Consider the force components in the x and y directions.

$$\text{Elemental force } \Delta x \text{ in } x \text{ direction} = p\delta s \sin\theta = p\delta y$$

$$\text{Total force in } x \text{ direction, } X = \oint p\,\mathrm{d}y$$

We define a two-dimensional force coefficient

$$C_x = \frac{X}{\frac{1}{2}\rho v^2 c} = \frac{\oint p\,\mathrm{d}y}{\frac{1}{2}\rho v^2 c}$$

i.e.

$$C_x = \oint \frac{p}{\frac{1}{2}\rho v^2}\,\mathrm{d}\left(\frac{y}{c}\right) \tag{8.1}$$

Similarly

$$C_y = \oint \frac{p}{\frac{1}{2}\rho v^2}\,\mathrm{d}\left(\frac{x}{c}\right) \tag{8.2}$$

We define a pressure coefficient $C_p = p/\frac{1}{2}\rho v^2$. Thus

$$C_x = \oint C_p\,\mathrm{d}(y/c) \tag{8.3}$$

$$C_y = \oint C_p\,\mathrm{d}(x/c) \tag{8.4}$$

These expressions for C_x and C_y can be evaluated from graphs which take the form of close contours.

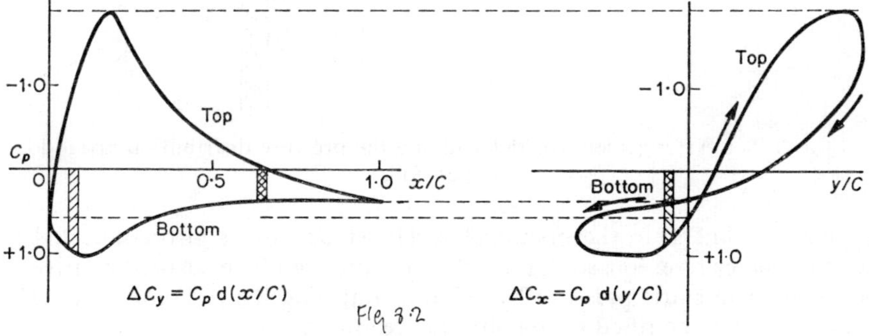

FIGURE 8.2 Pressure distributions over aerofoil.

The resultant value of ΔC_x at this particular value of y_L/C is given by the shaded area inside the closed contour.

Similar reasoning applied to elemental strips on the top surface will show that the net ΔC_x from the closed loop for the top surface gives a negative value to ΔC_x, as drawn.

In aerodynamic theory it is usual to express the resultant force in terms of lift and drag components, which are respectively normal and parallel to the free-stream direction. Hence

$$C_L = C_y \cos \alpha - C_x \sin \alpha \tag{8.5}$$

$$C_D = C_y \sin \alpha + C_x \cos \alpha \tag{8.6}$$

Note that C_x can be a negative quantity but C_D is *always* positive.

Experimental Details

Test Equipment. The aerofoil to be tested has a number of static pressure tappings in the top and bottom surface near the mid-span position, and these are connected to a multitube manometer as shown in Figure 8.3. A pitot-static tube is situated in the working section and the total and static pressure

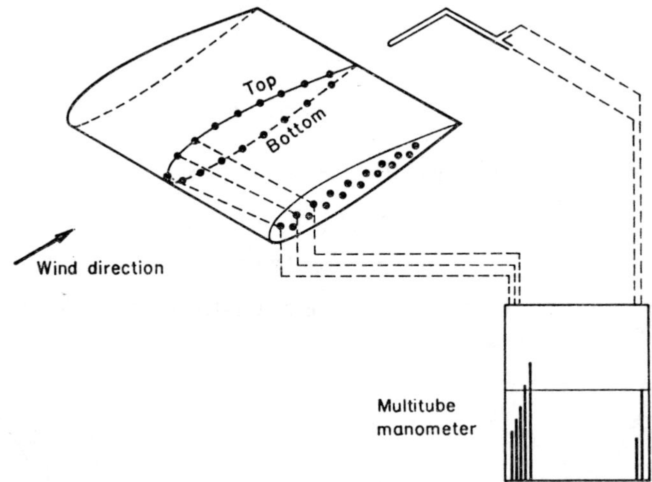

FIGURE 8.3 Arrangement for determining the pressure distribution around an aerofoil.

tappings are linked to the manometer. The static tube is also connected to the manometer reservoir so that all the pressures can be measured relative to the free-stream static pressure by suitably adjusting the reservoir level. The manometer may be tilted to magnify the readings.

Provision is made to allow the angle of incidence to be set accurately using an inclinometer.

Procedure. Set the aerofoil at a small negative angle of incidence. Ensure that the manometer is adjusted so that all the tubes are level. Start the fan motor and adjust the motor to give the required running speed for the tunnel. When the flow is steady, note the manometer readings. Note room temperature and pressure.

Repeat the above procedure for the complete range of incidences. Tabulate, for each angle of incidence, the value of $(p/\frac{1}{2}\rho v^2)$ for each pressure-tube position and, from the given values of x/c and y/c, plot graphs of $(p/\frac{1}{2}\rho v^2)$ against x/c and y/c and determine the values of C_x and C_y. Hence evaluate C_L and C_D and plot graphs of C_L against incidence, C_D against incidence, C_L against C_D and C_L/C_D against incidence.

Note the speed and the Reynolds' number and the Mach number of the test where

$$\text{Reynolds' number} = \frac{\rho v C}{\mu} \qquad \text{(dimensionless)}$$

where

$$\begin{aligned} \rho &= \text{density} && (\text{kg m}^{-3}) \\ v &= \text{velocity} && (\text{m s}^{-1}) \\ c &= \text{chord length} && (\text{m}) \\ \mu &= \text{viscosity} && (\text{kg m}^{-1}\,\text{s}^{-1}) \\ \text{Mach number} &= v/a && \text{(dimensionless)} \end{aligned}$$

and

$$a = \text{speed of sound} \qquad (\text{m s}^{-1})$$

Note that in each pressure plot there will be one point at which $p/(\frac{1}{2}\rho v^2) = +1{\cdot}0$, corresponding to the local stagnation point: the value of $p/(\frac{1}{2}\rho v^2)$ can never exceed $+1{\cdot}0$, although it can be less than $-1{\cdot}0$.

When plotting the pressures, difficulty will possibly be encountered with the y/c graphs because of the very rapid changes occurring. In these circumstances it will be helpful to refer to the corresponding x/c plot where the change in $p/(\frac{1}{2}\rho v^2)$ is more obvious and so additional points may be interpolated for the y/c plot. Errors in the evaluation of C_D can usually be traced to incorrectly drawn graphs.

Discussion

Outline the physical reasons for the observed shapes of the pressure plots and describe how these shapes change as the incidence increases. Compare the part played by the top and bottom surface in producing the lift force as the incidence increases. Explain why the drag force is small at low incidence but increases very rapidly at high incidence.

Comment on the shape of the C_L *v.* α, C_D *v.* α, C_L/C_D *v.* α and C_L *v.* C_D graphs and tabulate the following values:

(*a*) the slope of the C_L *v.* α graph,
(*b*) the incidence for which $C_L = 0$,
(*c*) the maximum value of C_L and the incidence at which it occurs,

(*d*) the minimum value of C_D and the incidence at which it occurs,
(*e*) the maximum value of C_L/C_D and the incidence at which it occurs.

Explain how the above graphs would be affected by the inclusion of the tangential shear forces.

Discuss possible errors in this test.

9

Flow Metering for Large Open Channels

Experiment 5 considered flow metering in closed ducts running full of fluid, but in many applications the liquid passes through large open channels for which it is necessary to employ a different type of flowmeter. In these instances it is essential that the flowmeter is capable of indicating over a wide range of flow rates and depths of fluid. Since debris can be deposited in an open channel, the flowmeter must not be too susceptible to blockage and must also be sturdily constructed.

The most commonly used flowmeters for this type of flow are (*a*) rectangular and vee-notches and (*b*) broad-crested weirs. Unlike the vee-notch, the rectangular notch does not always present the same geometrical shape to the flow, and is therefore not considered in this experiment.

Theory

(a) The vee-notch

Consider a vee-notch situated in an open channel such that the upstream water level is at a height H above the apex of the notch, as shown in Figure 9.1. Let b be the width of an elementary strip $\mathrm{d}h$ at a depth h below the still-water level.

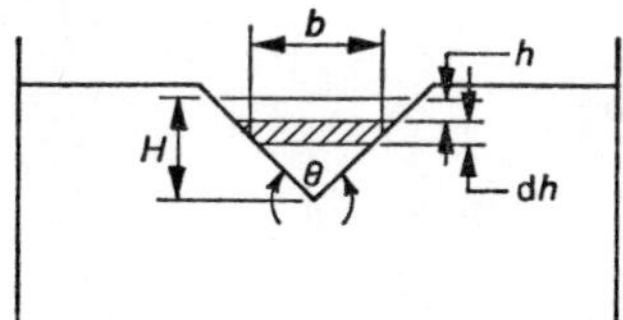

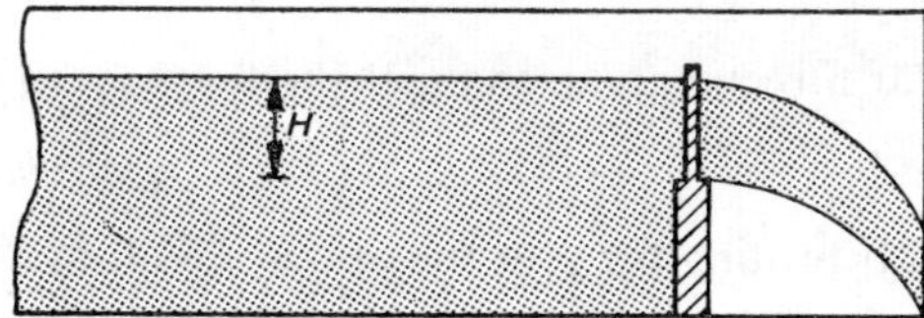

FIGURE 9.1 The vee-notch.

Assuming that the pressure is atmospheric when the liquid is passing through the notch and that the notch is running full to the height of the still-water level we have

$$\text{velocity at the elementary strip} = (2gh)^{\frac{1}{2}}$$
$$\text{discharge at the elementary strip} = b\,\mathrm{d}h(2gh)^{\frac{1}{2}}$$

Thus

$$\text{total discharge} = \int_0^H b(2g)^{\frac{1}{2}}h^{\frac{1}{2}}\mathrm{d}h$$

For a vee-notch,

$$b = 2 \tan \tfrac{1}{2}\theta(H - h)$$

Hence

$$\text{theoretical total discharge} = \int_0^H 2 \tan \tfrac{1}{2}\theta(2g)^{\frac{1}{2}}(H - h)h^{\frac{1}{2}}\,\mathrm{d}h$$

$$= 8(2g)^{\frac{1}{2}} \tan \tfrac{1}{2}\theta H^{5/2}/15 \qquad (9.1)$$

In practice neither of the above assumptions are correct and an empirical discharge coefficient, C_d, is introduced to give the actual discharge. Thus

$$\text{actual discharge} = C_d \times (\text{theoretical discharge})$$

$$= C_d \,.\, 8(2g)^{\frac{1}{2}} \tan \tfrac{1}{2}\theta \,.\, H^{5/2}/15 \qquad (9.2)$$

The actual value of C_d is dependent on the vee-notch angle but for a given notch it is constant in the working range of the notch.

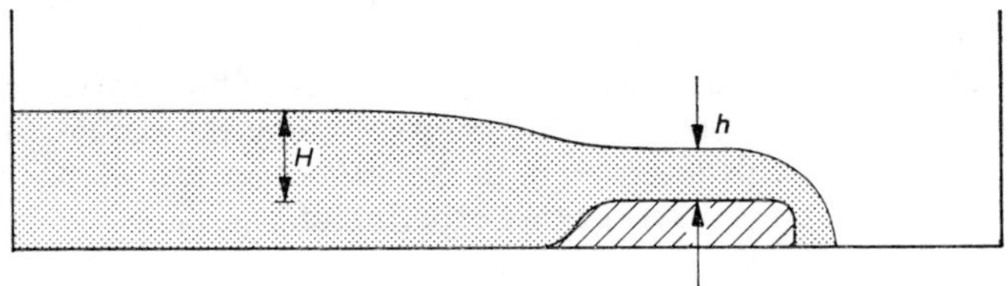

FIGURE 9.2 The broad-crested weir.

(b) The broad-crested weir

For flow in an open channel, the energy equation gives

$$H = h + v^2/2g \qquad (9.3)$$

For a rectangular channel the flow per unit width,

$$q = vh$$

Substituting for v in the above equation (9.3) gives

$$q^2 = 2g(Hh^2 - h^3) \qquad (9.4)$$

For maximum flow,

$$\mathrm{d}(q^2)/\mathrm{d}h = 0$$

and therefore

$$h = \tfrac{2}{3}H = h_c \qquad (9.5)$$

This is the critical depth of flow at which the volume rate of flow is a maximum.

Thus, provided that flow is at the critical depth condition, a single measurement of energy upstream is sufficient to gauge the flow rate. Under this condition

$$q = \tfrac{2}{3}H[\tfrac{2}{3}gH]^{\frac{1}{2}}$$

and
$$Q = qb = 1{\cdot}71\ bH^{3/2} \tag{9.6}$$

where Q is the flow rate in $m^3\ S^{-1}$ and b is the width of the channel in m.

In practice the volume flow will be less than the theoretical value and a weir constant is derived empirically.

Experimental Details

Test Equipment. The test equipment consists of a water channel in which a vee-notch or a broad-crested weir can be installed. Water is pumped from a sump through a venturi meter and into the settling chamber at the channel entrance. The chamber contains wire mesh screens and stones to act as a baffle and reduce turbulence in the channel flow. The height of the upstream still-water level is measured with a hook gauge.

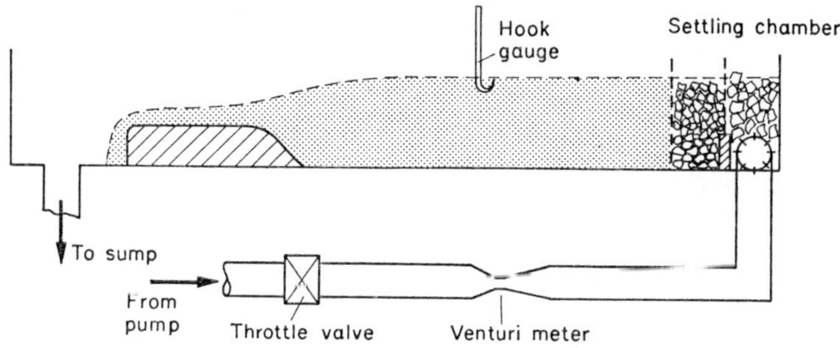

FIGURE 9.3 Open-channel flowmeter.

Procedure. With the vee-notch fitted in the channel, start the pump with the control valve fully closed and then gradually open the valve to give the maximum head required. When the flow has settled take the upstream water level reading relative to the notch apex and the venturi meter reading. By adjusting the control valve repeat the above procedure for several flow rates.

Remove the vee-notch and place the broad-crested weir in the channel. Repeat the above procedure taking the upstream water level reading relative to the top of the weir.

For each set of results plot a calibration curve of volume flow rate Q versus head of water H. Also plot log Q versus log H and find the slope and intercept of the line. As will be seen

$$Q = kH^n$$

describes the performance of both flow meters, and therefore

$$\log Q = \log k + n \log H \tag{9.7}$$

Thus the slope and intercept equal n and $\log k$ respectively.

Discussion

Compare the theoretical and experimental values of k and n for each flow meter and evaluate the discharge coefficient.

Discuss why the discharge coefficients for the two meters should be different and indicate what effects are dominant in each case.

Discuss the relevance of ignoring the velocity of approach by using the uniform water level as the total energy level.

Comment on the accuracy of each meter at low flow rates, and hence discuss the range of suitability of each meter for a given channel.

10

Pelton Wheel—Impulse Turbine

Useful work can be extracted from hydraulic energy in a variety of ways depending upon the required efficiency of operation and the acceptable degree of complexity in design. The two basic types of turbine are impulse and reaction machines. In the impulse turbine the total head of fluid is converted into kinetic energy before the fluid reaches the rotating part of the turbine, whilst for the reaction turbine only a portion of the total head is converted in this way.

The operation of a reaction turbine is similar to that of a centrifugal pump which will be examined in experiment 11. In this experiment we consider a type of impulse turbine called a Pelton wheel. It has a very high efficiency and is suitable for heads up to 1500 m.

Theory

An impulse turbine depends upon a free jet of fluid being turned through an angle θ by a series of moving blades. Consider one such blade as shown in Figure 10.1, where the inlet velocity is v_1, blade speed u, supply head H,

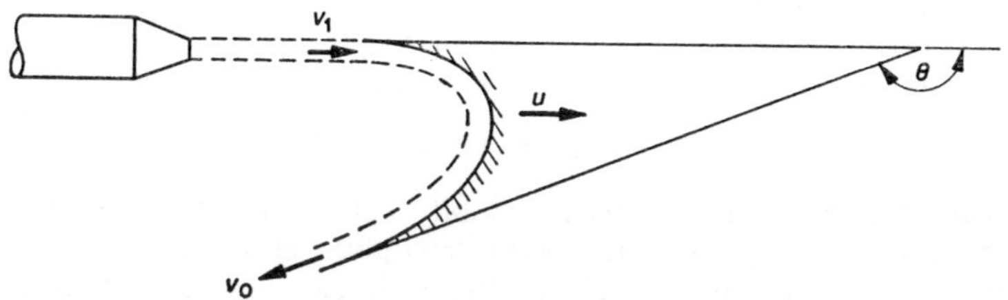

FIGURE 10.1 Jet deflection through angle θ.

nozzle velocity coefficient C_v, mass flow rate $\dot{m}$, blade angle θ, velocity at outlet v_0, relative velocity at outlet V_0 and relative velocity at inlet V_1, and where $V_0 = kV_1$, and k being the blade velocity coefficient.

FIGURE 10.2 Velocity triangles.

Change in relative velocity $= -kV_1 \cos(180° - \theta) - V_1$

$$\overrightarrow{\text{force on fluid}} = -\dot{m}(v_1 - u)[1 + k \cos(180° - \theta)] \quad (10.1)$$

$$\overrightarrow{\text{force on blade}} = \dot{m}(v_1 - u)[1 + k \cos(180° - \theta)] \quad (10.2)$$

Thus

$$\text{rate of doing work} = \text{force} \times \text{distance/time}$$
$$= \dot{m}(v_1 - u) \,.\, u[1 + k \cos(180° - \theta)] \quad (10.3)$$

For maximum power,

$$\frac{\mathrm{d}}{\mathrm{d}u}(\text{rate of doing work}) = 0$$

Hence

$$u = v_1/2. \quad (10.4)$$

This means that theoretically the blade speed for maximum power is half the jet speed. The theoretical maximum power is thus

$$\frac{\dot{m}v_1^2}{4}[1 + k \cos(180° - \theta)]$$

The power supplied to the wheel is $\dot{m}gH$. Hence

$$\text{hydraulic efficiency} = \frac{(v_1 - u) \,.\, u[1 + k \cos(180° - \theta)]}{gH} \quad (10.5)$$

But $v_1 = c_v[2gH]^{\frac{1}{2}}$ and therefore $gH = v_1^2/2C_v^2$. Hence

$$\text{hydraulic efficiency} = \frac{(v_1 - u) \,.\, u[1 + k \cos(180° - \theta)]}{v_1^2/2C_v^2} \quad (10.6)$$

So at maximum power (i.e. when $u = v_1/2$) the efficiency is given by

$$\tfrac{1}{2}C_v^2[1 + k \cos(180° - \theta)] \quad (10.7)$$

Thus for C_v and k approximately equal to unity and using a blade angle of approximately 180°, the maximum attainable efficiency is close to 100 per cent. In practice, for a Pelton Wheel efficiencies around 90 per cent are obtainable.

Where the power output is measured by a mechanical brake the overall efficiency can be quoted as

$$\left(\frac{\text{brake power}}{\text{water power}}\right)$$

Experimental Details

Test Equipment. A centrifugal pump draws water from a sump and delivers it through a venturi meter to the control valve at the turbine entry as shown in

Figure 10.3. The turbine consists of a number of buckets suitably shaped to turn the flow and the shaft of the turbine has a friction brake to measure the

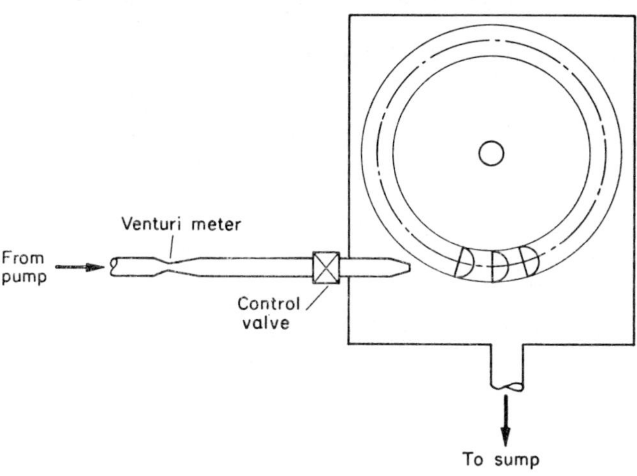

FIGURE 10.3 Pelton wheel turbine.

work output. The brake load is applied by a series of weights on a loading lever and the speed of rotation of the shaft is measured with a tachometer. A stroboscope may be used to study the flow path on to and away from the bucket and to illustrate the varying angle of departure as the rotational speed changes.

Procedure. Check that the discharge valve from the pump is fully closed and then start up the pump. Open this valve slowly, allowing the turbine to reach running speed and then adjust the pressure at the inlet to the turbine using the control valve and by-pass valve. When the turbine is running at a constant speed note the venturi meter reading, the brake load, and the rotational speed. The brake load is varied from zero at maximum r.p.m. to maximum load at approximately one-quarter maximum r.p.m., taking about eight sets of measurements in all. Repeat the above procedure for two other supply pressures. Tabulate the speed, brake load, power, volume flow and water energy for each supply pressure. For each supply pressure, plot graphs of power and efficiency against u/v_1.

Discussion

Consider the variation of power and efficiency with the speed ratio u/v_1 comparing the point of maximum efficiency with the theoretical position for $u/v_1 = 0{\cdot}5$.

Comment on the variation of maximum power with supply pressure and volume flow rate.

Compare the brake energy with the theoretical blade energy and explain the difference in terms of friction losses, etc., on the bucket, and friction and windage losses on the mechanical system.

Indicate the type and possible magnitude of errors in this test.

11

Centrifugal Pump Characteristics

Pressure loss in pipelines has many causes but in all cases energy is required in order to move fluids through the pipe. Sometimes potential energy may be used, i.e. flow can take place from a high position to a low level, but in the majority of cases a pump is used to supply the energy required.

There are many types of pump available, e.g. axial flow, radial flow, mixed flow, displacement, roller pumps, etc., but the commonest type used is the radial-flow or centrifugal pump.

Theory

All centrifugal pumps take in fluid near the centre and pass it outwards. Consider the inlet and outlet velocity triangles for a typical impeller vane as shown in Figure 11.1. Assume that the fluid enters the impeller with an absolute velocity in the radial direction. We have

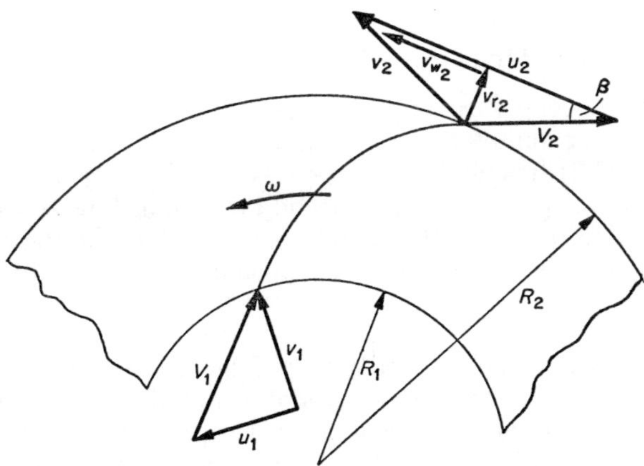

FIGURE 11.1 The action of a typical impeller vane.

Inlet Triangle

v_1 = absolute velocity
u_1 = vane speed = $R_1\omega$
V_1 = relative velocity

Outlet Triangle

v_2 = absolute velocity
u_2 = vane speed = $R_2\omega$
V_2 = relative velocity
v_{w_2} = whirl component of velocity, v_2
v_{r_2} = radial component of velocity, v_2

and ω = rotational speed (rad/sec).

Since the flow is radial at inlet there is no whirl component (i.e. in the direction tangential to the radius of the impeller) at that point. The radial velocities are given directly by the continuity equation. The work done can be found from the whirl component of velocity at outlet.

$$\text{Tangential momentum at outlet} = \dot{m}v_{w_2}$$
$$\text{Tangential moment of momentum} = \dot{m}v_{w_2}R_2$$

From the momentum equation,

$$\begin{aligned}\text{Torque on impeller} &= \text{change in moment of momentum}\\ &= \dot{m}v_{w_2}R_2\\ \text{Work done} &= \text{torque} \times \text{angular velocity}\\ &= \dot{m}v_{w_2}R_2\omega\\ &= \dot{m}v_{w_2}u_2\end{aligned}$$

Hence

$$\text{power required by impeller} = \rho Q v_{w_2}u_2 \qquad (11.1)$$
$$\text{power supplied} = \rho Q v_{w_2}u_2 + \text{losses} \qquad (11.2)$$

where the losses may be mechanical leakage or disc friction losses.

Experimental Details

Test Equipment. A centrifugal pump, driven by a variable speed electric motor, draws water from a sump and passes it through a venturi meter in the return circuit (Figure 11.2). Vacuum and over-pressure gauges are fitted immediately before and after the pump respectively and the volume flow is controlled by a valve. The suction pipe has a larger internal diameter than the delivery pipe. The power absorbed by the motor is calculated from a voltmeter and an ammeter.

Procedure. Check that the control valve is fully closed and that the pump is properly primed with water. Start the motor and allow it to become steady at the required test speed; take readings of the speed and pressure before and after the pump, and of the electrical voltage and current. Partially open the control valve and allow the flow to become steady, ensuring that the speed is kept constant; note the new readings and note also the venturi meter reading.

Repeat the above procedure for a range of flow rates and the whole procedure for a different speed.

Tabulate volume flow, pressure rise Δp, across the pump, speed, power and efficiency. Plot graphs of pressure rise, power and efficiency against volume flow for each speed N (rev/min).

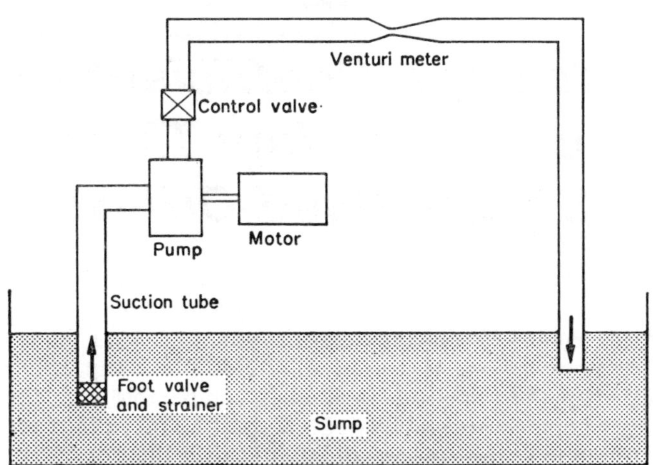

FIGURE 11.2 Centrifugal pump test rig.

Draw graphs of $\Delta p/\rho N^2D^2$, and η against Q/ND^3, where D = pump diameter, plotting the results for both speeds on the same graph. From the characteristic plot find the flow rate for maximum efficiency and calculate the radial velocity v_{r_2} at the outlet. From the geometry and the speed, calculate the vane speed u_2 at the outlet, and hence draw the theoretical outlet triangle assuming that the flow leaves the vane at an outlet angle β. Making an allowance for the losses, calculate the actual whirl component of velocity and superimpose the actual outlet triangle on the theoretical triangle.

Discussion

Comment upon the shape of the dimensional pump characteristics and on the condition for maximum efficiency.

Show by dimensional analysis that $\Delta p/\rho N^2D^2$ is a function of Q/ND^3 and find the corresponding non-dimensional power parameter. What is the effect of using the non-dimensional form of plotting? Discuss the dependence of the characteristics of a given pump on the speed of operation and suggest reasons why the characteristics for a given pump are often quoted relative to 1000 rev/min.

Comment on the differences between the theoretical and actual outlet triangles. Quote the ratio of actual whirl component to theoretical whirl component and state how this is likely to change as the flow rate changes. Why is the suction pipe larger than the delivery pipe?

12

Discharge Coefficient for Compressible Flow Through a Sharp-edged Orifice

The theory developed in previous experiments applies to fluids in the incompressible flow region. However, all fluids are compressible to some extent, and when the velocity of a fluid is comparable with the speed of sound for that fluid, compressibility effects must be taken into account. The parameter which indicates compressibility in this context is called the Mach number M, and is defined as

$$M = \frac{\text{(fluid velocity)}}{\text{(local velocity of sound in fluid)}} = \frac{v}{a} \tag{12.1}$$

In air, compressibility effects are negligible below $M = 0{\cdot}4$. In this simple test the effect of compressibility on the flow through a sharp-edged orifice is investigated.

Theory

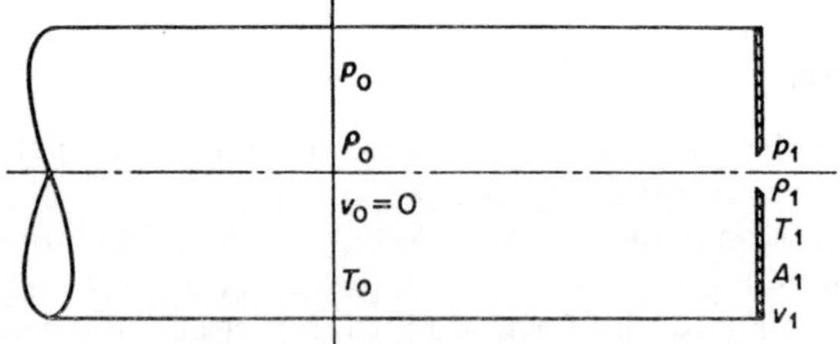

FIGURE 12.1 A sharp-edged orifice.

Consider the flow through a sharp-edged orifice across which there is a pressure differential $(p_0 - p_1)$. The volume upstream of the orifice is large enough to be considered as a reservoir. We have

Continuity equation: $\rho_1 A_1 v_1 = \dot{m}$ (12.2)

Energy equation: $$\frac{v_1^2}{2} = \frac{\gamma}{\gamma - 1}\left[\frac{p_0}{\rho_0} - \frac{p_1}{\rho_1}\right] \tag{12.3}$$

Isentropic law: $$\frac{\rho_0}{\rho_1} = \left(\frac{p_0}{p_1}\right)^{1/\gamma} \tag{12.4}$$

Equation of state: $$\frac{p_0}{\rho_0 T_0} = \frac{p_1}{\rho_1 T_1} = R \tag{12.5}$$

where R is the gas constant. From the energy equation

$$v_1^2 = \frac{2\gamma}{\gamma - 1}\frac{p_0}{\rho_0}\left[1 - \frac{p_1}{p_0}\frac{\rho_0}{\rho_1}\right]$$

This becomes by the use of the equation of state and the isentropic law

$$v_1^2 = \frac{2\gamma}{\gamma - 1} RT_0\left[1 - \left(\frac{p_1}{p_0}\right)^{\frac{1-\gamma}{\gamma}}\right]$$

$$v_1 = \left\{\frac{2\gamma}{\gamma - 1} RT_0\left[1 - \left(\frac{p_1}{p_0}\right)^{\frac{1-\gamma}{\gamma}}\right]\right\}^{\frac{1}{2}}$$

Substituting this in the continuity equation eventually leads to

$$\frac{\dot{m}(RT_0)^{\frac{1}{2}}}{A_1 p_0} = \left\{\frac{2\gamma}{\gamma - 1}\left(\frac{p_1}{p_0}\right)^{2/\gamma}\left[1 - \left(\frac{p_1}{p_0}\right)^{\frac{1-\gamma}{\gamma}}\right]\right\}^{\frac{1}{2}} \tag{12.6}$$

Thus the theoretical non-dimensional mass flow parameter $\dot{m}(RT_0)^{\frac{1}{2}}/A_1 p_0$ is a function of γ and the pressure ratio p_1/p_0.

In practice the mass flow will be less than the theoretical value and a discharge coefficient C_d is introduced into the equation, so that it becomes

$$\left[\frac{\dot{m}(RT_0)^{\frac{1}{2}}}{A_1 p_0}\right]_{\text{actual}} = C_d\left\{\frac{2\gamma}{\gamma - 1}\left(\frac{p_1}{p_0}\right)^{2/\gamma}\left[1 - \left(\frac{p_1}{p_0}\right)^{\frac{1-\gamma}{\gamma}}\right]\right\}^{\frac{1}{2}} \tag{12.7}$$

This coefficient C_d depends upon the Mach number, which is a function of the pressure ratio.

If the theoretical non-dimensional mass flow parameter is plotted against pressure ratio, the curve obtained (Figure 12.2) indicates that a maximum mass flow should occur at a critical pressure ratio of 0·528 for air with $\gamma = 1.41$. However, the portion of the curve to the left of the critical pressure ratio does not exist in practice since the flow through the orifice theoretically reaches sonic velocity at the critical pressure ratio. The orifice becomes 'choked', the mass flow parameter remaining constant for pressure ratios less than the critical.

In practice the curve will lie below the theoretical since C_d is less than unity and varies from the incompressible value of approximately 0·6, tending towards 1 as the pressure ratio is reduced, i.e. as the Mach number is increased.

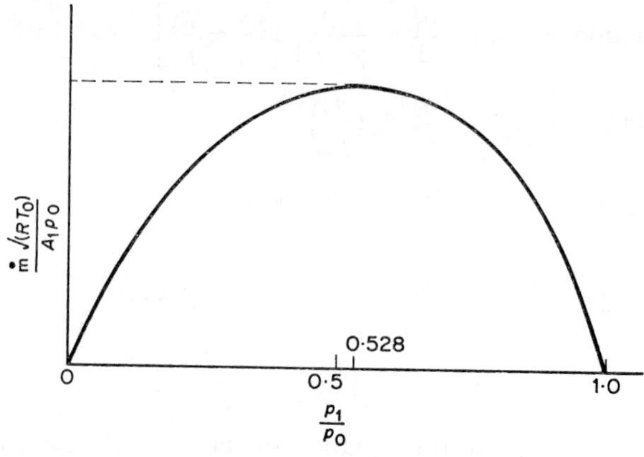

FIGURE 12.2 Theoretical curve of non-dimensional mass-flow parameter versus pressure.

Experimental Details

Test Equipment. A large-diameter pipe leads air from a large reservoir of compressed air (which is both clean and dry) through an automatic control valve and a large orifice meter to the small orifice plate at the pipe exit (Figure 12.3). The pressure drop across the large orifice is measured with a high-pressure liquid manometer and the pressure and temperature upstream

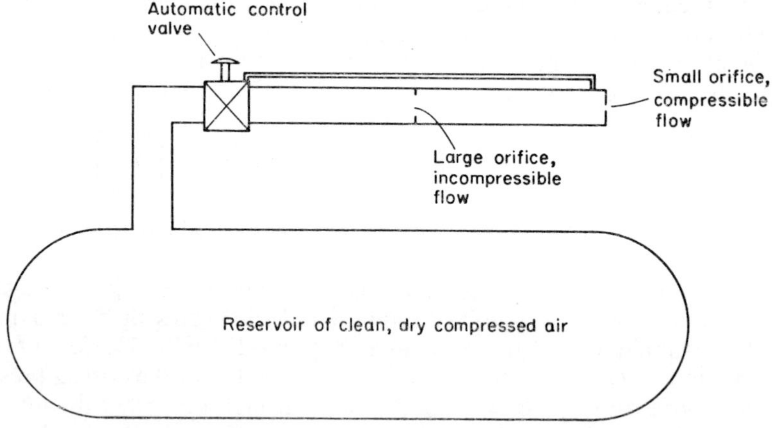

FIGURE 12.3 Compressible flow through an orifice.

of the small orifice are given by direct reading instruments. The automatic control valve can be set to maintain the downstream pressure constant at any selected value even though the reservoir pressure decreases during the test.

Procedure. Charge the reservoir to its maximum working pressure and set the required pressure on the automatic control valve. Activate the control valve, and when the flow has become steady take readings of temperature and pressure at the small orifice and the pressure drop across the large orifice. Recharge the reservoir and repeat the above procedure for a series of set pressures.

For each run calculate the pressure ratio, the density in the main pipe and hence the actual mass flow by using the incompressible flow equation for the large orifice. Tabulate and plot the actual and theoretical values of $\dot{m}(RT_0)^{\frac{1}{2}}/A_1p_0$ and thus calculate and plot the discharge coefficient C_d.

Discussion

Show by differentiation of the mass flow equation that the theoretical maximum value of the non-dimensional mass flow parameter occurs at the critical pressure ratio 0·528 for $\gamma = 1{\cdot}4$. Show from the relationship

$$p_0/p_1 = [1 + \tfrac{1}{2}(\gamma - 1)M_1^2]^{\gamma/\gamma-1} \tag{12.8}$$

that this corresponds to flow at $M = 1{\cdot}0$.

Discuss the variation of C_d with pressure ratio and give reasons why the actual maximum mass flow parameter does not occur at the theoretical critical pressure ratio.

Assuming that the flow becomes sonic at the minimum cross-sectional area for pressure ratios less than 0·528, describe with the aid of sketches how the jet shape will change for pressure ratios less than 0·528.

Comment on the differencc in mass flow $\dot{m}$ obtained by varying the pressure ratio by (*a*) decreasing the downstream pressure p_1 and (*b*) increasing the upstream pressure p_0.

Part II

Thermodynamics

Introduction

One of the aims of a laboratory course is to stimulate the student's interest by demonstrating applications in practice of the principles and theory he is learning. This objective is, perhaps, more difficult to achieve in thermodynamics than in any other branch of engineering, for in many of the experiments which normally form part of an undergraduate course, the way in which thermodynamic principles are applied is not very obvious. For instance in a performance test on an internal combustion engine, which is a necessary part of current undergraduate courses, thermodynamics often features explicitly only fairly remotely in a comparison of actual and Carnot cycle efficiencies. This feature makes some contribution to the relative unpopularity of thermodynamics courses, though doubtless the main basis of the lack of enthusiasm is the difficulty of understanding concepts such as entropy. All this is rather unfortunate for there is no doubt that thermodynamics holds a commanding position as the most fundamental single topic and also provides, for those who choose to use it, a most effective bridge between applied and pure science offering, for engineering students, a very good opportunity of gaining a greater insight into the underlying physical principles governing the world around us.

In the experiments described below, which attempt to provide a balanced laboratory course, the choice of experiments has been made on the basis that they should permit as directly as possible the emphasis of thermodynamic theory, e.g. by discussing entropy as an active parameter in experiment 21 and by careful discussion of the concept of temperature.

The latter concept is so readily capable of intuitive interpretation that its essential relation to equilibrium states is often overlooked. Few students are greatly perturbed by the concept of a temperature which changes with time nor by the use of thermodynamic functions in fluid mechanics equations where possibly both temperature gradients, thermodynamic functions and time variations appear in the same equation. A good example of this is the energy equation which can be written in the form

$$\rho T \frac{\mathrm{D}s}{\mathrm{D}t} = \Phi + \operatorname{div}(K \operatorname{grad} T)$$

(Here T denotes temperature, Φ is the dissipation function, s is entropy, t is time, K is thermal conductivity, ρ is density and $\mathrm{D}/\mathrm{D}t$ denotes the total derivative with respect to time).

The fact that such equations can, in practice, be used effectively, is an interesting reflection on our concepts of general and local thermodynamic equilibrium. Nevertheless students should not become so used to using equations of this type that they lose sight of the essential limitations and restrictions which are implicit in the use of thermodynamic concepts in situations which involve states which are clearly not truly equilibrium states.

Heat transfer, even if steady, is essentially a non-equilibrium process. For this reason, though heat transfer often features as part of an undergraduate thermodynamics course, no experiments on heat transfer have been included in this section. However a classical experiment on radiation which involves both some heat transfer and also a change of temperature with time has been included since it provides an interesting illustration of the application of thermodynamic reasoning in an unusual context.

At the other end of the scale, some introductory experiments have been described which require no basic knowledge other than the ideal gas laws. These are intended to help bridge the gap between school A-level courses and undergraduate thermodynamics courses. Experiment 1, on an air compressor, is particularly useful in this respect as it is a good introduction to many of the concepts used in general in relation to the reciprocating motion of a piston in a cylinder.

Most of the experiments described can be set up in the average thermodynamics laboratory and can easily be extended, if facilities permit, to become more interesting, more varied and possibly more difficult. It is hoped that at least some of them will suggest lines of thought which may help in the construction of undergraduate laboratory courses in thermodynamics.

13

The Reciprocating Air Compressor

The principle on which the action of a simple reciprocating air compressor is based is straightforward and its practical application in an actual compressor is not difficult to understand. A compressor experiment is a very convenient way of introducing at an early stage in an undergraduate thermodynamics laboratory, indicator diagrams and the concepts of mean effective pressure and cycles of operations (though not, of course, thermodynamic cycles since the air which is compressed does not return to its initial state after each cycle of operations).

Theory

The principal measurement in this experiment is of the rate at which the pressure in the storage vessel increases for a series of constant discharge pressures from the compressor. From this several useful inferences can be drawn.

Let V_0 be the volume of the storage vessel,

$\dot{V}$ be the volume of air passing through the compressor in unit time measured at outlet pressure P_f and temperature T_f,

ρ_f be the density of the air leaving the cylinder of the compressor at P_f and T_f.

Then the mass flow of air to the storage vessel in unit time is given by

$$\dot{M} = \rho_f \dot{V}$$

If the compressor motor is operating at constant speed then $\dot{M}$ is constant, and the mass of air which has entered V_0 in time t is $\dot{M}t$. Assuming V_0 was initially filled with air at atmospheric pressure P_i and temperature T_i then the total mass in V_0 after time t is

$$(\dot{M}t + \rho_i V_0)$$

where, if air is assumed to behave as an ideal gas with a gas constant R per unit mass,

$$\rho_i = P_i/RT$$

Hence the density ρ_t of air in V_0 after time t has elapsed is given by

$$\rho_t = \rho_i + \frac{\dot{M}t}{V_0} = \frac{P_t}{RT_t}$$

where P_t and T_t denote respectively the pressure and the temperature of the air in V_0 at time t. Hence

$$\frac{P_t}{T_t} = \frac{P_i}{T_i} + \left[\frac{\dot{M}R}{V_0}\right]t \tag{13.1}$$

Hence if P_t/T_t is plotted against time t the slope will give the value of $\dot{M}R/V_0$ and the intercept on the P_t/T_t axis the value of P_i/T_i. The numerical value of the intercept can easily be checked by direct measurement and gives some indication of the accuracy of the measurements. The slope can be used to obtain an experimental value for the *volumetric efficiency* of the compressor.

Since the air contained in the clearance volume V_c (Figure 13.1) must

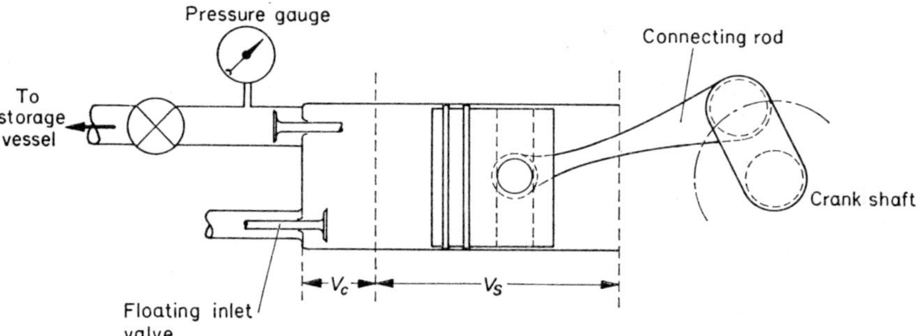

FIGURE 13.1

expand until its pressure falls to atmospheric before any air can enter through the inlet valve, the actual volume of atmospheric air entering the cylinder of the compressor is less than the swept volume V_S. The volumetric efficiency η_{vol} of the compressor is a measure of this effect and is defined by the relation

$$\eta_{vol} = \frac{\text{volume of air induced per cycle}}{\text{swept volume}}$$

where the volume induced is measured at inlet pressure P_i and temperature T_i.

$$\eta_{vol} = \frac{\text{(volume induced/cycle) (number of cycles/s)}}{\text{(swept volume)(number of cycles/s)}}$$

$$= \frac{\text{volume induced per second at } P_i \text{ and } T_i}{\text{swept volume} \times \text{number of cycles per second}}$$

If the compressor driving shaft makes N revolutions per second,

$$\eta_{\text{vol}} = \frac{\dot{M}}{\rho_i} \cdot \frac{1}{NV_S}$$

$$= \frac{\dot{M}RT_i}{P_i} \cdot \frac{1}{NV_S}$$

$$= V_0 \frac{\text{(Slope obtained from equation 13.1)}}{\text{(intercept obtained from equation 13.1)}} \frac{1}{NV_S} \qquad (13.2)$$

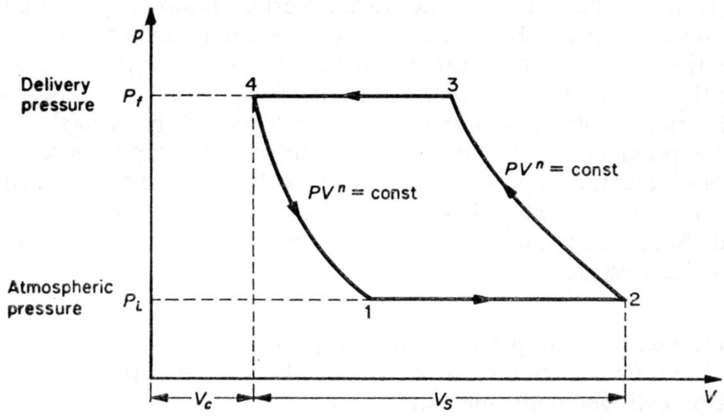

FIGURE 13.2

A value for η_{vol} can be obtained directly if it is assumed that the compressor is reversible and that both expansion and compression strokes obey the law PV^n = constant, where $0 < n < \gamma$. Referring to Figure 13.2,

$$\eta_{\text{vol}} = \frac{V_2 - V_1}{V_S} = \frac{V_S + V_c - V_1}{V_S}$$

$$= 1 + \frac{V_c}{V_S} - \frac{V_c}{V_S} \cdot \frac{V_1}{V_c}$$

Now

$$P_1 V_1^n = P_4 V_4^n = P_4 V_c^n$$

Hence

$$\eta_{\text{vol}} = 1 - \frac{V_c}{V_S} \left\{ \left[\frac{P_4}{P_1} \right]^{1/n} - 1 \right\} \qquad (13.3)$$

Comparison of equations (13.2) and (13.3) for given values of P_4 enables the value of n to be estimated (assuming the compressor to be reversible) from a graph of $\log(1 - \eta_{\text{vol}} + (V_c/V_S))$ *versus* $\log(P_4/P_1)$.

Experimental Details

Test Equipment. This experiment needs only a simple single-stage reciprocating compressor, but can be carried out with any commercial compressor provided it has easy access to the outlet from the first stage. The outlet valve from the cylinder should be connected directly to an air storage vessel by a pipe fitted with a pressure gauge and with a valve at the inlet to the air storage vessel. (Figure 13.1) The gauge and the valve will make it possible to keep constant the pressure at which air leaves the cylinder while the pressure in the storage vessel increases, so that both inlet and outlet conditions for the compressor are constant. The storage vessel itself should also be fitted with a pressure gauge and an outlet valve. Normal Bourdon pressure gauges will be quite adequate for this experiment. A thermometer is required to measure the temperature of the air in the storage vessel, and another to measure the air inlet temperature. The shaft of the compressor motor should be fitted with a revolution counter and a stop watch will be needed. Facilities should be provided on the cylinder for the measurements necessary for plotting an indicator diagram. Details of the cylinder bore and the length of the piston stroke and also the volume of the air storage vessel should be provided. Access to a barometer will be necessary in order to find absolute values of the pressures.

Procedure. Measure the pressure and temperature in the air storage vessel at approximately one minute intervals with the discharge pressure from the compressor cylinder controlled in turn at about five discrete values, e.g. 3, 4, 5, 6 and 7 bars. Before beginning each set of measurements the air storage vessel should be at atmospheric pressure with its outlet valve closed and with the delivery valve from the compressor also closed. The compressor motor can then be started. When the pressure in the tube connecting the compressor to the storage vessel (Figure 13.1) is at the desired level, open the compressor delivery valve to admit air to the storage vessel. Throughout the experiment adjust the delivery valve manually as required to maintain a constant value of discharge pressure from the compressor.

While the pressure in the storage vessel is being raised, take readings at approximately one minute intervals of the revolution counter on the crankshaft of the compressor. Measure also the temperature of the air at the inlet to the compressor and read the laboratory barometer. At least one indicator diagram should be obtained during each set of readings.

Follow the same procedure for each of the chosen delivery pressures and in each case plot the value of the absolute pressure P_t in the storage vessel at time t divided by its absolute temperature T_t against time. Find the slope of this line by the method of zero sum and from it the mass of air passing through the compressor in unit time (using equation (13.1)) and also the volumetric efficiency of the compressor (using equation (13.2)). Obtain a value for the latter also from the indicator diagrams by drawing on each a line representing atmospheric pressure, as shown in Figure 13.3. Then $\eta_{vol} = BD/AC$. Note that BD does not coincide with the part of the indicator diagram representing the position of the cycle where the pressure is lowest since there is usually a small pressure drop across the inlet valve; also the points C and D do not usually coincide since the resistance to the flow of air through the inlet valve is sufficient to make the actual time taken for the valves to open and close a small but detectable portion of the total time taken for one complete cycle.

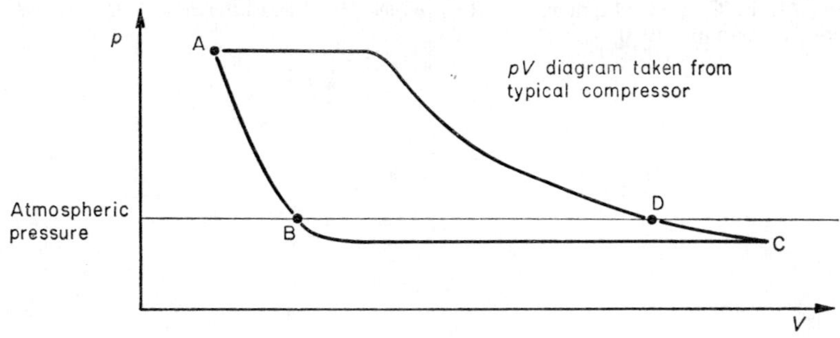

FIGURE 13.3

A quantity of interest in describing the performance of a compressor is the rate of volume flow measured at inlet conditions P_i and T_i. This is termed the *free air delivery* (F.A.D.) of the compressor. Clearly

$$\text{F.A.D.} = \frac{\dot{M}}{\rho_i} = V_i$$

$$= V_0 \frac{T_i}{P_i} \left(\text{slope of } \frac{P_t}{T_t} \text{ } vs. \text{ } t\right)$$

Plot graphs, and comment on them, of the following quantities against the delivery pressure P_4 (Figure 13.2) as base:

(*a*) F.A.D.,
(*b*) η_{vol} from equation (13.2), and
(*c*) η_{vol} from the indicator diagrams.

Further Experiments

As an extension of the basic experiment, the compressor can be driven by an electric motor. The difference in the power the compressor absorbs when idling at a given speed and when operating normally at the same speed can then be measured. These readings enable an estimate to be made of the work done by the compressor which can be compared with the values obtained from the indicator diagram and with those obtained using the value of n obtained from equation (13.3) and applying ideal gas laws. The concept of mean effective pressure is easily introduced at this stage.

As a development of the simple measurement of η_{vol}, ways of improving its value lead naturally to the two-stage compressor and the effects of inter-stage cooling, on which further experiments can easily be made if required.

Reference

Rogers, G. F. C. and Mayhew, M. R. *Engineering Thermodynamics: Work and Heat Transfer*, Longmans, 1970.

14

Relation Between Scales of Temperature

Temperature or, more precisely, temperature difference is such a simple intuitive concept that its correct thermodynamic meaning is frequently overlooked. In particular the fact that temperature can be defined thermodynamically only for equilibrium states is seldom emphasized and the concept of a temperature changing with time as, for example, while plotting a cooling curve, is only infrequently deemed to require some qualification.

The least ambiguous physical definition of temperature is that it is the property which is common to all equilibrium states of a system which lie on an isotherm. Here, to be rigorous, an isotherm must be defined using the zeroth law of thermodynamics as the locus of all points representing on a property diagram (such as a *p–v* diagram for a gas) all the states of one system which are in thermal equilibrium with a single state of another system. A precise and more straightforward definition may be obtained from the usual equation expressing the combined first and second laws of thermodynamics, namely

$$T\,\mathrm{d}s = \mathrm{d}u + p\,\mathrm{d}v \tag{14.1}$$

It follows that

$$\begin{aligned}\mathrm{d}s &= \frac{1}{T}\,\mathrm{d}u + \frac{p}{T}\,\mathrm{d}v \\ &\equiv \left(\frac{\partial s}{\partial u}\right)_v \mathrm{d}u + \left(\frac{\partial s}{\partial v}\right)_u \mathrm{d}v\end{aligned}$$

and hence

$$\frac{1}{T} \equiv \left(\frac{\partial s}{\partial u}\right)_v \tag{14.2}$$

In this definition it is not immediately obvious that temperature is defined only for equilibrium states but, of course, it is implicit in the use of equation (14.1) which is valid only for the thermodynamics of equilibrium states.

To measure temperature any physical property which varies with temperature can be used to construct a thermometer. There is an implicit assumption here too in that it is taken for granted that any practical thermometer will be of sufficiently low thermal capacity not to perturb the temperature which is being measured and of sufficiently low thermal inertia to come quickly into equilibrium with the system whose temperature is to be measured. Now it would be unreasonable to expect any two arbitrarily chosen physical properties to vary with temperature in the same way. For any system of thermometry to be viable it is essential that any two arbitrarily chosen thermometers each using the variation with temperature of a different physical property should nevertheless give the same value for the temperature of a given constant-temperature heat reservoir. The only temperature scale which is independent of the physical properties of thermometric materials is the absolute thermodynamic scale, which depends only on measurement of quantities of heat. This provides a universal reference scale against which all other thermometers can be calibrated, directly or indirectly. For any practical thermometer, utilizing the variation with temperature of some physical property of a substance, a scale of temperature is defined in terms of the values of that property at certain thermometric fixed points such as the boiling and freezing points of pure water at s.t.p. The purpose of the present experiment is to relate such individual temperature scales.

Theory

Consider a thermometer which detects temperature changes in terms of the variation of some physical property X of a substance, e.g. X could be the resistance of a wire, the pressure of a gas whose volume was kept constant, a thermal e.m.f., or the length of a column of mercury. It is usual to define a scale of temperature in terms of measured values of X making use of the fixed values X_0 of X at the freezing point of water at s.t.p. and the value X_{100} of X at the boiling point of water at s.t.p. It is further assumed that the interval between these two fixed points is divided into 100 equal units (degrees). Then rises of temperature are *defined* as being in the same ratio as increases in the value of the property X by the relation

$$\frac{\theta}{100} = \frac{X_\theta - X_0}{X_{100} - X_0} \tag{14.3}$$

where X_θ denotes the value of X at temperature θ measured on this scale of temperature. This, of course, defines a linear scale of temperature which will be valid only for temperatures measured in terms of the variation of X. The actual variation of X with thermodynamic temperature may well not be linear.

Suppose that above the ice point, X is related to thermodynamic temperature by the equation

$$X = A + BT + CT^2 \tag{14.4}$$

where A, B and C are constants and where T is the difference between true thermodynamic temperature and the thermodynamic temperature at the ice point.

From equation (14.3)

$$\frac{\theta}{100} = \frac{(A + BT + CT^2) - (A + B.0 + C.0^2)}{(A + B.100 + C.100^2) - (A + B.0 + C.0^2)}$$

since at the ice and steam points both θ and T have the same numerical values. Hence

$$\theta = \frac{BT + CT^2}{B + 100C}$$

$$T - \theta = T - \frac{BT + CT^2}{B + 100C}$$

$$= \frac{10^4C}{(B + 100C)}\left[\frac{T}{100} - \left(\frac{T}{100}\right)^2\right]$$

This may be written in the form

$$T - \theta = \delta\left[\left(\frac{T}{100}\right)^2 - \frac{T}{100}\right] \tag{14.5}$$

where

$$\delta = -\frac{10^4C}{B + 100C} \tag{14.6}$$

The quantity δ can then be used to correct the temperature θ (based on some physical property X) to the absolute temperature ($T + T_{\text{ice point}}$).

Experimental Details

Test Equipment. Since both T and θ have, by definition, exactly the same values at the ice and steam points equation (14.5) can be used only with temperatures different from 0 °C and 100 °C. Measurement of T and θ at a single temperature is sufficient to evaluate δ and use of a fixed point such as the boiling point of sulphur (444·60 °C) would be ideal, except that this is a rather inconveniently high temperature. Instead it is recommended that a good laboratory furnace is used to provide a temperature in the region of 200 °C which is measured by a laboratory sub-standard platinum resistance thermometer. There will probably be a temperature gradient in such a furnace, which provides an opportunity for students to overcome a common experimental difficulty.

Three thermometers are recommended for the experiment:

(*a*) a constant volume gas thermometer,
(*b*) a nickel resistance thermometer (the resistance characteristic of nickel being markedly non-linear), and
(*c*) a thermocouple.

The resistance of the platinum and nickel resistance thermometers can be measured either with a Wheatstone bridge or by using a good quality potentiometer to compare the resistance of the thermometer with that of a standard resistance (e.g. a 1 ohm standard) connected in series with the thermometer.

As a potentiometer is necessary for measuring the thermocouple e.m.f., it is probably simpler to use a potentiometer throughout.

Finally constant temperature baths at 0 °C and 100 °C are needed to establish the temperature scales for the constant volume gas thermometer, the nickel resistance thermometer and the thermocouple, using equation (14.3). (Note that for the thermocouple $X_0 = 0$ and in equation (14.4) $A = 0$). Also a barometer is needed to find the absolute pressure in the constant volume thermometer.

Procedure. Establish a temperature scale for each of the thermometers under investigation by measuring in each case the value of the appropriate thermometric property X at the ice point and again at the steam point. (The calibrated sub-standard thermometer indicates the actual temperature of the two heat reservoirs used.) Usually these two temperatures will differ by some quantity different from 100 °C. In this case if X_1 and X_2 denote the property values at the ice and steam points respectively and the measured temperature interval is Y °C then equation (14.3) should be modified to

$$\frac{\theta}{Y} = \frac{X_\theta - X_1}{X_2 - X_1}$$

Use each thermometer in turn to measure the temperature θ_X of the furnace in terms of the variation of the chosen thermometric property X. At the same time measure the absolute temperature T of the furnace with the laboratory sub-standard thermometer.

Find δ_X for each of the thermometers used from equation (14.5), and construct graphs of θ_X *vs.* T for any desired temperature range, using the measured value of δ_X in each case to calculate the value of θ_X corresponding to any arbitrarily chosen values of T. This should show clearly the difference between the various scales of temperature.

Further Experiments

A profitable extension which can be made from this experiment is to compare the various ways of making thermometers. The thermocouple is the simplest case to deal with in a normal laboratory and it demonstrates the effects of thermal cycling of the hot junction. A drift in the readings is usually detectable, sometimes for as many as 100 cycles. Various ways of making up cold junctions can also be tried. A method found very convenient in practice is to join the two thermocouple wires (e.g. chromel and alumel) to two copper leads and to surround both junctions and the leads from them with paraffin wax. When the wax has set it can easily be inserted as a block in an ice water mixture in a Dewar flask. This eliminates any difficulties due to short circuiting the thermocouple leads by the cold junction liquid and due to the small vertical temperature gradient often encountered in a Dewar (assuming the two wire junctions are in the same horizontal plane). A little time is, of course, needed for the paraffin wax to attain a uniform temperature of 0 °C.

15

Measurement of the ratio γ of the principal specific heats for air

The measurement of γ by Rüchhardt's method and its modification by Rinkel provide very versatile student experiments. The principle is easy to understand, the compressibility of air is demonstrated neatly and effectively, good values of γ are obtained with very simple apparatus and with different degrees of accuracy for each of three possible methods of making the measurement, and finally several sophisticated modifications and extensions are possible.

Theory

The principle of the method is to compress air in a closed volume with a ball which is a close fit inside a glass tube and which acts essentially as a piston, the rate of change of momentum of the ball after a small displacement providing the necessary force. Since this small force is not capable of producing a large change in volume the arrangement chosen is that illustrated in Figure 15.1 with the ball making an effective seal in the tube D which is connected to the large volume V. The small volume change resulting from forces of the order of magnitude of the weight of the ball will appear as appreciable vertical movement of the ball in the tube D.

Suppose the ball is at an equilibrium position at the level XY with air trapped in the volume V together with part of the tube D, the whole of the trapped volume being denoted by V. This can easily be arranged by inserting the ball in the tube D and carefully allowing air to escape through the valve Q. Suppose Cartesian coordinate axes are set up with XY lying in the xy-plane. If the ball is displaced upwards from XY by some force, z will increase and the enclosed volume will also increase so that $\mathrm{d}V$ and $\mathrm{d}z$ will both be positive. Similarly for downward displacement both $\mathrm{d}V$ and $\mathrm{d}z$ are negative. If then the ball is given a vertical displacement z the corresponding volume change $\mathrm{d}V$ will be of the same sign as z and is given by

$$\mathrm{d}V = zA$$

where A is the cross-sectional area of the tube D.

The corresponding pressure change dP will, however, be of opposite sign to z. If the displacement z is imagined to be due to a small force F acting on the ball then

$$\mathrm{d}P = F/A$$

Since F is of opposite sign to z, it is a restoring force and the ball will oscillate about its mean position. These oscillations will be sufficiently rapid for the changes in the pressure and volume of the enclosed air to be regarded as adiabatic; furthermore, since the amplitude of the oscillations is small, the various states through which the air passes in one complete oscillation are effectively equilibrium states and the whole process approximates very closely to a quasistatic adiabatic process. In this case

$$PV^{\gamma} = \text{constant}$$

and it follows that

$$P \,.\, \gamma V^{\gamma-1}\,\mathrm{d}V + V^{\gamma}\,\mathrm{d}P = 0.$$

Now we have shown that d$V = zA$ and d$P = F/A$, and so

$$\frac{\mathrm{d}P}{\mathrm{d}V} = -\frac{\gamma P}{V} = \frac{F}{A^2 z}$$

$$F = -\frac{\gamma P A^2}{V} z \qquad (15.1)$$

Since the force F is proportional to the displacement z of the ball and of opposite sign, the motion of the ball will be simple harmonic and its period T will be given by

$$T = 2\pi(m/\mu)^{\frac{1}{2}}$$

where m is the mass of the ball and μ is the magnitude of the restoring force per unit displacement, i.e.

$$T = 2\pi\left[\frac{mV}{\gamma PA^2}\right]^{\frac{1}{2}}$$

$$\gamma = \frac{4\pi^2 mV}{A^2PT^2} \qquad (15.2)$$

In this experiment m, V and A are apparatus constants, T is the period of oscillation of the ball and P the pressure in the volume V when the ball is at rest. Clearly P will be given by the relation

$$P = P_A + \frac{mg}{A}$$

where P_A denotes atmospheric pressure. Hence all the quantities in equation (15.2) can be evaluated and a value of γ obtained.

In practice, of course, there will be friction between the ball and the glass tube so the motion of the ball will be a damped simple-harmonic motion. This damping is small and will not seriously affect the period of the oscillation, so the simple analysis neglecting friction is not invalidated.

An interesting modification of this experiment was introduced by Rinkel. He arranged for the ball to be held at the mouth of the tube D and allowed it to drop from rest into the tube, having first arranged for the initial pressure in the volume V to be exactly atmospheric. Suppose under these conditions the ball falls a distance L before it is brought to rest and its direction of motion reversed. The change in gravitational potential energy of the ball is $(-mgL)$ and this must equal the work done in compressing the air in V from atmospheric pressure to its final value. It follows that

$$-mgL = \int_0^{-L} F\,dz$$

$$= \int_0^{-L} -\frac{\gamma PA^2}{V} z\,dz \qquad \text{(from equation (15.1))}$$

$$= \frac{-\gamma PA^2}{V}\frac{L^2}{2}$$

$$\therefore \qquad \gamma = \frac{2mgV}{PA^2L} \qquad (15.3)$$

Using this method L has to be measured instead of T, and this can usually be done with greater accuracy than the measurement of the period since the oscillatory motion is damped out after a comparatively small number of oscillations. Also, only the first power of L is needed in equation (15.3), whereas the second power of T appears in equation (15.2), thus doubling any experimental error in the measurement of T itself.

The two methods can, of course, be combined with advantage, measuring first L and then the period of the subsequent motion. This is the recommended method, preferably using the technique described on p. 68.

Experimental Details

Test Equipment. The feasibility of the experiment depends entirely on obtaining a glass tube of the required length which permits the ball to fall freely with a minimum of friction and yet with a clearance small enough to provide a good seal—requirements which conflict. A 16 mm steel ball is recommended for use with a 50 cm length of glass tubing mounted vertically in the neck of a glass flask V about 6 litres in volume (Figure 15.1). A 16 mm ball normally gives a nearly perfect seal in a length of standard 16 mm bore tubing but the friction is too high to permit sufficiently free movement of the ball. For the experiment to be successful the tube must be chosen carefully to permit the ball to descend smoothly at a rate of about 1 cm s^{-1} when the tube is vertical and its lower end is sealed. (It will probably be necessary to ask glass manufacturers to produce specially a set of suitable tubes.) With a combination of ball and tube which meets this requirement and a vessel V of about 6 litre capacity containing air at atmospheric pressure, a 16 mm ball falling from rest will drop a distance L of about 38 cm before its direction of motion is reversed. The subsequent oscillation will consist of about 12 cycles at a frequency of 1 Hz before friction finally brings the ball to rest. Both the value of L and the period T of the subsequent oscillation must be measured.

The initial pressure inside V can be checked with the manometer M and reduced to atmospheric if necessary by opening the valve Q. To ensure that the ball is dropped from rest a small metal cup fitted with a release pin is mounted on the end of the glass tube as shown. Without such an arrangement it is difficult to ensure reproducible release conditions. It is essential to clean the ball in acetone before each measurement and to handle it with tongs—otherwise the ball tends to stick in the tube.

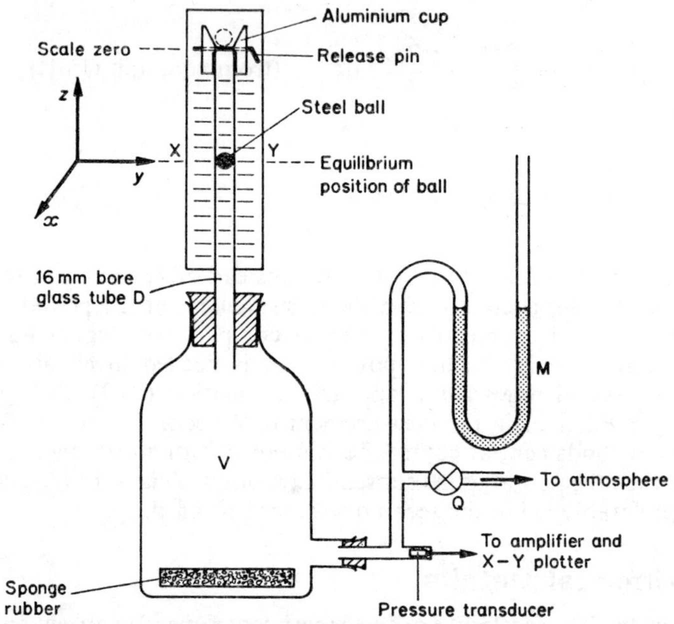

FIGURE 15.1

A scale is attached to the tube with its zero coinciding with the position of the release pin so that the distance L can be measured directly. The period T of the oscillations can be measured directly too, using a stop-watch.

More sophisticated and instructive measurements can be made by recording, as a function of time, on an X–Y plotter, the output from a pressure transducer connected to the vessel V. The maximum amplitude of the pressure fluctuation is about 1·5 per cent of atmospheric pressure, and as its frequency is about 1 Hz it is easy to detect a fluctuation of this nature with a fairly simple pressure transducer. With such an arrangement a trace similar to that illustrated in Figure 15.2 will be obtained. This makes possible several improvements. It is now possible to obtain a value of the period T and hence of γ to $\pm\frac{1}{2}$ per cent, compared with ± 1 per cent by measuring L and ± 2 per cent by manual timing of the oscillations. The logarithmic decrement of the oscillation and the decrease in the mean level of the trace can both be measured making it possible to assess the damping due to friction and the loss of mass due to leakage. Finally by static calibration of the trace it is possible to use it to measure the maximum pressure reached in V after the ball has fallen a distance L. A static calibration using manometer M to measure the pressure is easily made with the aid of a bicycle pump connected to Q with a cork inserted in the end of the tube. It is then possible to measure directly the change in pressure corresponding to a volume decrease equal to the product of L and the cross sectional area of tube.

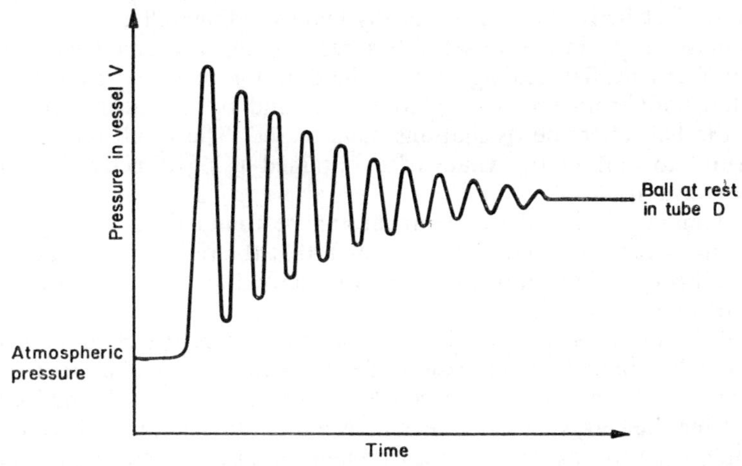

FIGURE 15.2

Procedure. Clamp the cup containing the release pin firmly to ensure that removal of the release pin does not bend the glass tube (which is long enough to be reasonably flexible). Check its operation using a steel ball which has previously been carefully cleaned with acetone and allowed to dry by evaporation. Adjust the pressure transducer and X–Y plotter amplifiers to give a trace approximately in the middle of the recording paper and make a few trial runs. At this stage, if required, calibrate the chart by removing the cup, sealing the tube with a cork and raising the pressure in V through the valve Q. Measure the pressure in V directly with the manometer M to calibrate manually the static deflection of the pen of the X–Y recorder.

Next replace the cup on the tube and adjust the scale so that its zero coincides with the top surface of the release pin (i.e. the bottom of the ball). Measure the distance L using a movable marker by adjusting the position of the marker until the bottom surface of the ball appears just to touch it before the ball begins to move upwards.

Measure L directly and T both directly and from the trace. Calculate values of γ and compare these values by the following methods:

(*a*) using PV^γ = constant directly, with measured values of the pressure increment following the dropping of the ball and calculating the corresponding volume change,
(*b*) using the two measured values of T,
(*c*) using the measured value of L.

The validity of the assumptions made in each case should also be discussed.

Further Experiments

Several extensions of the basic experiment are possible, none is easy to effect but all are instructive both theoretically and experimentally.

The mass of gas in the vessel V has been assumed constant whereas this is not in fact true. The leakage during the duration of an experiment can be estimated both from the X–Y plotter trace and by measuring the rate of fall of the ball after the oscillations have ceased. The result could be used to attempt to correct the value of γ obtained using directly the relation PV^γ = constant.

The damping during the experiment can be found from the logarithmic decrement of the trace and if it is assumed that the resistance to the motion of the ball is directly proportional to its velocity, the effect on the measured period can be estimated.

Finally it is in principle possible to measure temperature changes in V due to the nearly adiabatic compression of the gas in it. To make this measurement will require a very low inertia thermometer in good thermal contact with the enclosed air, such as, for example, a thin film resistance thermometer. It is interesting to attempt this measurement even if fraught with difficulty. This extension could well form the basis of a small student project.

References

Zemansky, M. W. *Heat and Thermodynamics* (London: McGraw–Hill, 1943.)
Frost, T. H. *Joblinglass*, April 1968, pp. 16–21.

16

Measurement of Steam Quality

Several ways can be devised for measuring the quality (or dryness fraction) of steam, with varying degrees of accuracy. Apart from mechanical separation of the water by some form of centrifuge, two methods, described here, are of thermodynamic interest as illustrations of the use which may be made of the property 'enthalpy'.

(a) Use of a Throttling Calorimeter

Theory

Suppose a line is carrying steam of quality q, where q is sufficiently close to unity for a throttling process to take the steam into the superheated region. This is illustrated in Figures 16.1 and 16.2, where in each case point 1 represents the initial state of the steam and point 2 the final superheated state.

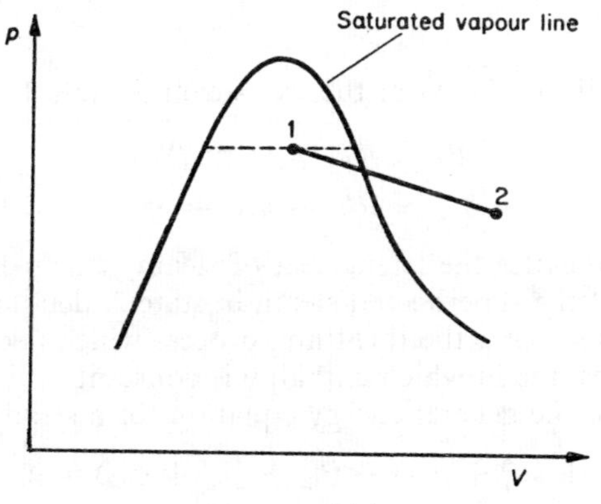

FIGURE 16.1

If p_1 and T_1, the pressure and temperature of the steam in state 1, are measured, steam tables will give the enthalpy h_{g1} of saturated vapour and h_{f1} of saturated liquid at p_1 and T_1. If the steam now undergoes a throttling process in a well lagged calorimeter which is so designed that the steam which emerges, at a pressure p_2 and a temperature T_2, is superheated then again the enthalpy h_2 of the vapour can be obtained from the superheated steam tables. If the steam then enters a condenser where it is condensed at a temperature T_3 and at atmospheric pressure p_3, its mass flow rate may be measured by weighing the condensate collected in a given time. These measurements, together with a measurement of the temperature rise in the cooling water in the condenser provide several ways of finding the initial quality of the steam.

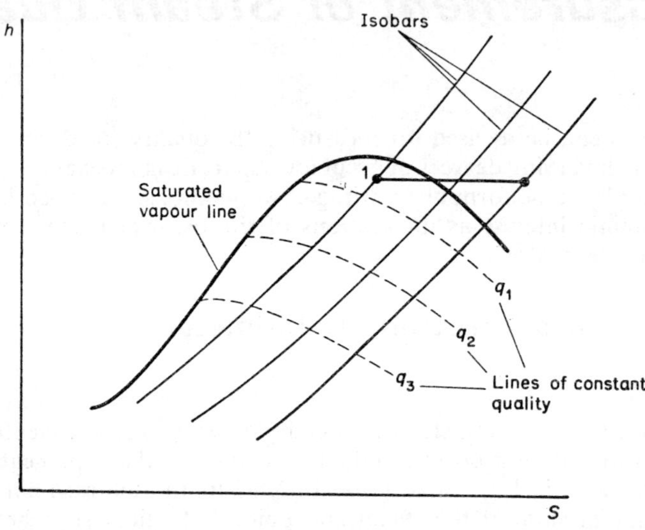

FIGURE 16.2

The initial specific enthalpy of the wet steam in state 1 is given by

$$h_1 = qh_{g1} + (1 - q)h_{f1} \tag{16.1}$$

$$= q(h_g - h_f)_1 + h_{f1} \tag{16.2}$$

where $(h_g - h_f)_1$ denotes the latent heat of steam at temperature T_1.

The enthalpy of the superheated steam in state 2, denoted by h_2, will have the same value as h_1 since the throttling process which takes the steam from state 1 to state 2 is one in which enthalpy is constant. This follows from the integrated form of the general energy equation for a steady flow process

$$(h_1 + \tfrac{1}{2}v_1^2 + gz_1) - (h_2 + \tfrac{1}{2}v_2^2 + gz_2) = w - q$$

where the notation is the usual one. In the throttling process kinetic and potential energy terms are negligible, no external work is done and no heat flow to or from the system occurs, leaving

$$(h_1 - h_2) = 0$$

Hence
$$q(h_g - h_f)_1 + h_{f1} = h_2 \tag{16.3}$$

All the enthalpies in equation (16.3) can be found from steam tables when p_1, T_1, p_2 and T_2 are known, and hence q may be found.

A further value may be found from consideration of what happens in the condenser, which is normally at atmospheric pressure p_3, at which pressure the saturation temperature of steam is T_3. The steam is first cooled at constant pressure p_3 until it has reached the saturation temperature T_3 when it begins to condense. If c_p denotes the mean specific heat of superheated steam in the temperature range T_2 to T_3, and h_{f3} the enthalpy of the water when all the steam has condensed at temperature T_3

$$h_2 - c_p(T_2 - T_3) - (h_g - h_f)_3 = h_{f3} \tag{16.4}$$

Since the superheated steam is cooled at constant pressure p_3

$$c_p(T_2 - T_3) = (h_2 - h_3)_{p_3} \tag{16.5}$$

where h_2 and h_3 denote specific enthalpies of superheated steam at the pressure p_3. Substituting in equation (16.4) and using equation (16.3),

$$q(h_g - h_f)_1 + h_{f1} = (h_2 - h_3)_{p_3} + (h_g - h_f)_3 + h_{f3} \tag{16.6}$$

Alternatively, use may be made of the heat balance in the condenser to write equation (16.6) in a different form. If M_0 is the mass flow rate of the condenser coolant water whose specific heat is c, and δT is the rise in its temperature between inlet and outlet, and if M is the mass flow rate of the condensate,

$$M_0 c\delta T = M[c_p(T_2 - T_3) + (h_g - h_f)_3] \tag{16.7}$$

Equations (16.3), (16.4) and (16.7) may now be combined to give

$$q(h_g - h_f)_1 + h_{f1} = h_{f3} + \frac{M_0 c\delta T}{M} \tag{16.8}$$

Finally, if the quality of the steam supply is not sufficiently close to unity for the throttling process to take it into the superheated region, its quality can be raised by using a mechanical steam separator mounted upstream of the throttling calorimeter. Suppose such a separator removes a mass m of water from the steam in unit time when conditions are steady. Then the actual dryness fraction q_1 in the steam line is given by

$$q_1 = \frac{qM}{(M + m)}$$

Experimental Details

Test Equipment. Some form of steam separator which permits measurement of the mass of water removed from the steam flow should be connected between a steam line and the throttling calorimeter illustrated in Figure 16.3. The calorimeter should be well insulated to make it reasonable to assume that very little heat flows in or out of the system (as was assumed in the theory). A thermometer and pressure gauge are fitted to the calorimeter as shown to measure p_2 and T_2. The throttling is provided by a small orifice in the calorimeter. After leaving the calorimeter the steam, now superheated, is passed to a condenser where it is completely condensed, and the mass flow rate M of the condensate should be measured, together with the pressure p_3 (or temperature T_3) at which condensation ensues. A scales and a stop-watch will be needed for this operation. Finally the scales, the stop-watch and two

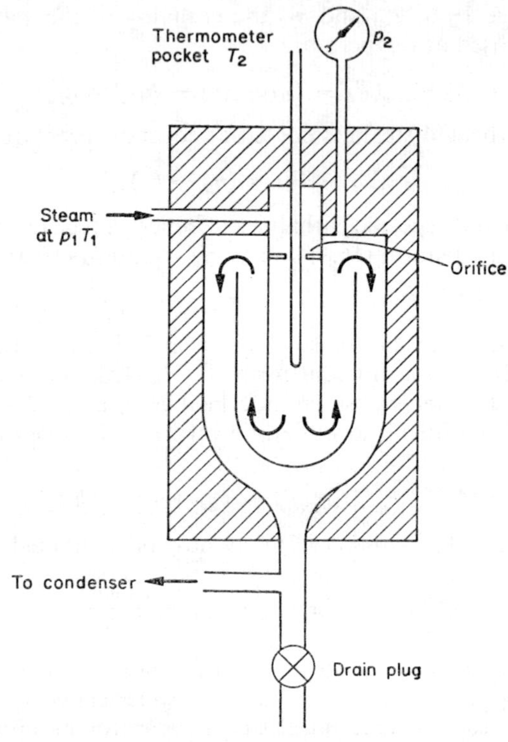

FIGURE 16.3

further thermometers are needed to measure the mass flow rate of cooling water in the condenser and its temperature rise.

Procedure. Before the experiment starts, allow steam to flow through the steam line and the throttling calorimeter for at least ten minutes to permit the steam line and the calorimeter to reach a state of thermal equilibrium—otherwise the quality of the steam flowing will vary considerably along the length of the line and in the calorimeter. When the whole apparatus is in thermal equilibrium with its immediate surroundings, drain off all water from the mechanical water separator, the throttling calorimeter and the condenser. Then, with the steam flow maintained at a steady value for a measured time interval, make the following measurements:

(i) the amount of water collected in the mechanical steam separator;
(ii) the pressure in the steam line;
(iii) the pressure and temperature in the calorimeter;
(iv) the pressure (or temperature) in the condenser;
(v) the flow rate of the condensate;
(vi) the flow rate of the condenser coolant; and
(vii) the temperature rise in the condenser coolant.

Repeat the whole procedure for at least one further measured time interval. The time intervals chosen should be as long as possible to increase experimental accuracy and in practice are probably determined by the capacity of the collectors of the condenser cooling water and condensate.

The steam quality q can be found using equations (16.3), (16.6) and (16.8).

(b) Injection of Steam into Water

Theory

Consider a closed system composed of several sections between which mass may flow but in which the total mass is constant. Each of the sections may separately exchange heat and work with its surroundings so that the generalized expression for conservation of energy in the whole system must involve a summation of terms, one for each section, of the form

$$\Delta(mu)_i = Q_i - W_i \tag{16.9}$$

where m is the mass in the section i, and u is its specific internal energy, both of which may change during the course of a process; Q_i is the heat flow across the boundary of the section; and W_i is the external work done by the fluid in the section. For n sections the generalized statement of the first law is

$$\sum_{i=1}^{n} \Delta(mu)_i = \sum_{i=1}^{n} Q_i - \sum_{i=1}^{n} W_i \tag{16.10}$$

Using the relation $h = u + pv$ this equation becomes

$$\sum_{i=1}^{n} \Delta(mh)_i = \sum_{i=1}^{n} Q_i - \sum_{i=1}^{n} W_i + \sum_{i=1}^{n} \Delta(mpv)_i \tag{16.11}$$

Now if the pressure in each section is constant (and only in this case)

$$\begin{aligned} W_i &= p_i\Delta(mv)_i \\ &\equiv \Delta(mpv)_i \qquad \text{where } p \text{ is constant} \end{aligned}$$

Equation (16.11) becomes

$$\sum_{i=1}^{n} \Delta(mh)_i = \sum_{i=1}^{n} Q_i \tag{16.12}$$

If each section is well insulated so that no heat crosses its boundary $Q_i = 0$ and equation (16.12) reduces to

$$\sum_{i=1}^{n} \Delta(mh)_i = 0 \tag{16.13}$$

i.e. the total enthalpy change of the entire system is zero if the pressure in each section remains constant. (There is no requirement for the pressure to be the same in each section.)

Suppose this result is applied to the system illustrated in Figure 16.4 and composed of two systems, one a large vessel A open to the atmosphere and

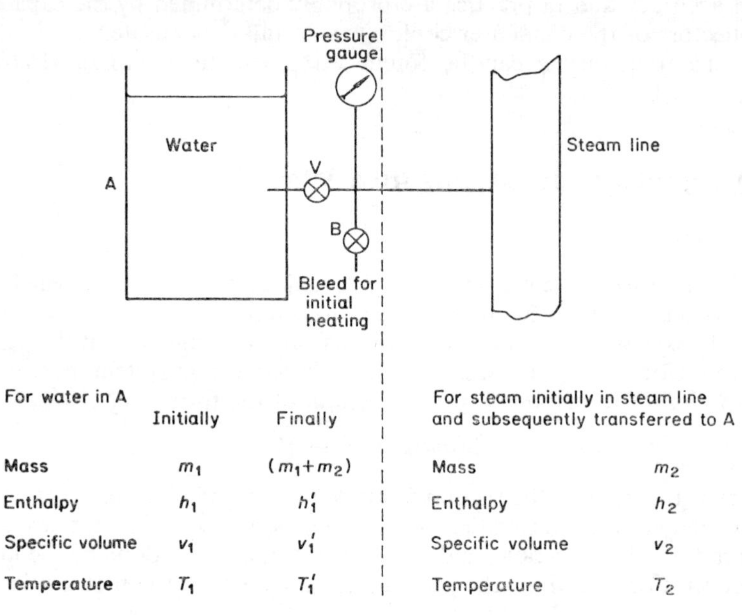

FIGURE 16.4

containing a mass m_1 of water at T_1 and the other an imaginary section containing a mass m_2 of steam in a steam line at a constant temperature T_2, which during the course of the process is transferred to the vessel A, raising its temperature to T_1'. Equation (16.13) applied to this case gives

$$m_1h_1 + m_2h_2 = (m_1 + m_2)h_1' \tag{16.14}$$

The values of m_1 and m_2 may be found by weighing and h_1 and h_1' using steam tables and the measured initial and final temperatures T_1 and T_1' of the water in A. It is then easy to find h_2, the specific enthalpy of the steam in the line. If q is the quality of the steam in the line at T_2,

$$h_2 = qh_g + (1 - q)h_f \tag{16.15}$$

where h_g and h_f are obtained from steam tables at a temperature T_2 and hence q may be found.

Experimental Details

Test Equipment. The apparatus is very simple, comprising merely a large vessel containing water with a connection from a steam line entering the vessel beneath the water surface. The vessel A, the steam line and the connection between them should be insulated. A large weighing scale is needed to measure the initial and final masses of water in A, and also two thermometers, one for the water temperature and one for the steam.

Procedure. Pass steam through the steam line (with valve V closed and B open) to warm it up, and meanwhile measure the mass m_1 of water in the vessel A. When the steam line and, as far as possible, the connection to A are at a steady temperature, close the valve B and open V to admit steam to A at such a rate that the steam is absorbed in the water. When steam has been passed into A for a long enough time for the mass of condensed steam to be determined with reasonable accuracy, close the valve V and measure the temperature of the water in A. Measure the mean temperature of the steam in the steam line during the course of the experiment and check that the steam pressure remains constant—this is one of the basic assumptions in the method. Find the quality of the steam from equation (16.15).

Further Experiment

The effect on the measured values of q obtained by both methods described of taking samples from several points in the steam line could be investigated.

References

1. Zemansky, M. W. and Van Ness, H. C. *Engineering Thermodynamics*, McGraw-Hill, 1966.
2. Rogers, G. F. C. and Mayhew, Y. R. *Engineering Thermodynamics: Work and Heat Transfer*, Longmans, 1970.

17

Change of Phase—The Clausius-Clapeyron Equation

An experiment illustrating the variation of saturation pressure of water with temperature can easily be constructed using quite simple apparatus. The method described is based on the fact that water boils at that temperature at which its vapour pressure equals the pressure of its immediate surroundings. If the external pressure is varied from about 0·1 atmosphere up to 1 atmosphere the boiling point varies from about 50 °C to 100 °C. By reducing the pressure in a water containing system a graph of saturation pressure *vs.* temperature can be obtained easily. The slope of the graph at any chosen temperature can be used, in conjunction with steam tables, to verify the Clausius-Clapeyron equation.

Theory

The specific value of the Gibbs function is defined in the usual notation by the equation

$$g = u - Ts + pv \tag{17.1}$$

Hence
$$dg = du - T\,ds - s\,dT + p\,dv + v\,dp$$

$$= v\,dp - s\,dT \tag{17.2}$$

(since $T\,ds = du + p\,dv$). Therefore, for any changes occurring at constant temperature and pressure, $dg = 0$.

If the change in question is one of phase, a mass dm of water may be supposed to change from liquid to vapour. If g_f denotes the specific value of the Gibbs function for water and g_g that for steam at the same pressure and temperature, then for the system as a whole,

$$dg = dm(g_f - g_g)$$

Since $dg = 0$ it follows that

$$g_f = g_g.$$

This relation will hold at any value of pressure and temperature though the actual values of g_f and g_g will not be independent of pressure and temperature. For a change dp in the pressure and dT in the temperature,

$$\mathrm{d}g_f = \left(\frac{\partial g_f}{\partial p}\right)_T \mathrm{d}p + \left(\frac{\partial g_f}{\partial T}\right)_p \mathrm{d}T \tag{17.3}$$

$$\mathrm{d}g_g = \left(\frac{\partial g_g}{\partial p}\right)_T \mathrm{d}p + \left(\frac{\partial g_g}{\partial T}\right)_p \mathrm{d}T \tag{17.4}$$

If $g_f = g_g$ at all values of p and T it must follow that

$$\mathrm{d}g_f = \mathrm{d}g_g$$

Now from equation (17.2), since g is an exact differential

$$\left(\frac{\partial g}{\partial p}\right)_T = v \quad \text{and} \quad \left(\frac{\partial g}{\partial T}\right)_p = -s$$

Using these relations in equations (17.3) and (17.4)

$$v_f\,\mathrm{d}p - s_f\,\mathrm{d}T = v_g\,\mathrm{d}p - s_g\,\mathrm{d}T$$

Hence

$$\frac{\mathrm{d}p}{\mathrm{d}T} = \frac{s_g - s_f}{(v_g - v_f)}$$

For a change of phase at constant pressure, dh = dq, so that at the temperature T_0,

$$T_0(s_g - s_f) = h_g - h_f$$

where h denotes specific enthalpy. Hence

$$\frac{\mathrm{d}p}{\mathrm{d}T} = \frac{h_g - h_f}{T_0(v_g - v_f)} \tag{17.5}$$

This is the Clausius-Clapeyron equation.

Experimental Details

Test Equipment. The vessel V (see Figure 17.1) should have a volume of several litres so that pressure fluctuations in the system are very small. The pressure in the system is reduced by an air-ballast rotary pump, which can be isolated from the system by the tap T, the pressure being measured either by a mercury manometer or by a Bourdon-type vacuum gauge. The water in the glass flask F should contain a few porcelain beads to promote smooth boiling and its temperature should be measured with a suitable thermometer—a mercury in glass thermometer is quite adequate. A barometer is also needed.

Procedure. With the vessel V open to the atmosphere, heat the water in the flask F until its temperature is about 50 °C. Connect the vacuum pump to the system and slowly reduce the pressure until the water in F begins to boil.

At this stage close the tap T and disconnect the vacuum pump. When the water is boiling steadily measure its temperature T and the pressure p in the system. Admit air to the system by opening the tap T until the pressure has increased by about 0·1 bar, and heat the water in F until it again begins to boil; then remove the source of heat. When the water is boiling steadily take readings of the pressure and temperature. Repeat the process at pressure intervals of about 0·1 bar until the pressure in the system is atmospheric.

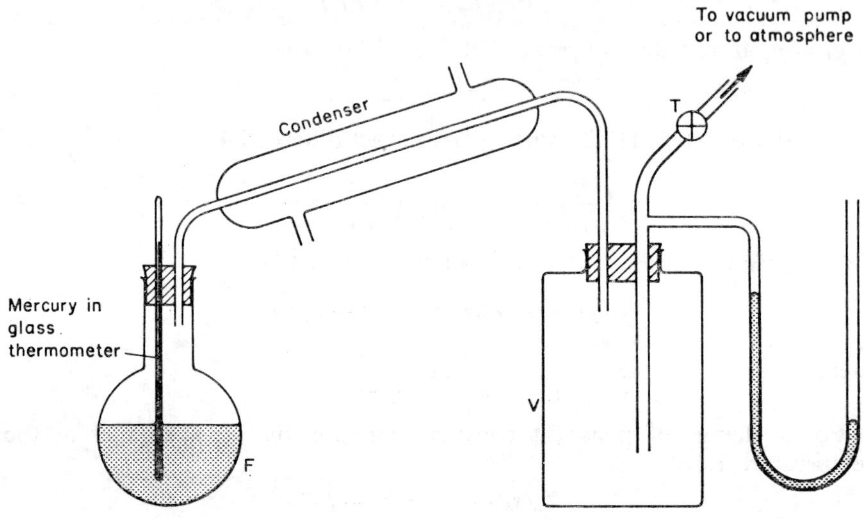

FIGURE 17.1

Plot a graph of p *versus* T, and measure the slope of the graph at some arbitrarily chosen value of the temperature T within the range covered by the experiment. Use this value of slope together with steam tables to verify the Clausius-Clapeyron equation. Repeat the procedure for several points on the graph.

Further Experiment

Instead of finding $(h_g - h_f)$ from steam tables a simple experiment could be performed to measure the latent heat of water in the range 50–100 °C.

18

The Stirling Cycle

An experiment with a heat engine operating in a manner approximating to the Stirling cycle offers several features of interest. (Nevertheless in an undergraduate laboratory context it can be little more than a demonstration experiment.) It is a cycle which, with an ideal regenerator, is capable of an overall efficiency equal to the Carnot efficiency and with a much larger mean effective pressure; it is a convenient introduction to the concept of regenerators and the problems of heat transfer; it can be used either for the conversion of heat into work as a simple heat engine or for operation as a refrigerator; and, finally, the cycle is of increasing interest because of its importance as a practical refrigeration cycle offering the greatest efficiency available in the temperature range 70 K to 170 K.

Theory

The basic cycle is a non-flow process consisting of two isothermal and two constant volume processes as illustrated in Figures 18.1 and 18.2. During

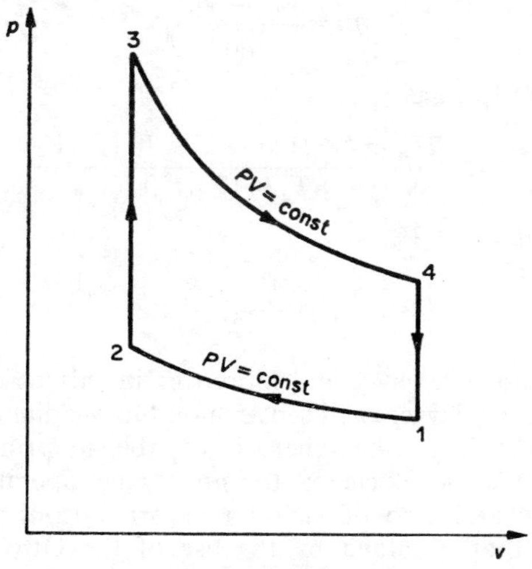

FIGURE 18.1

the constant volume processes no work is done but heat transfer will be necessary to cause the pressure changes. The efficiency of the cycle is given by the equation

$$\eta = \frac{\text{work output}}{\text{heat input}} = \frac{w_{34} - w_{12}}{q_{23} + q_{34}} \tag{18.1}$$

Since processes $1 \rightarrow 2$ and $3 \rightarrow 4$ are isothermal, q_{23} must equal q_{41} though these quantities of heat will be of opposite sign. Hence if the heat rejected

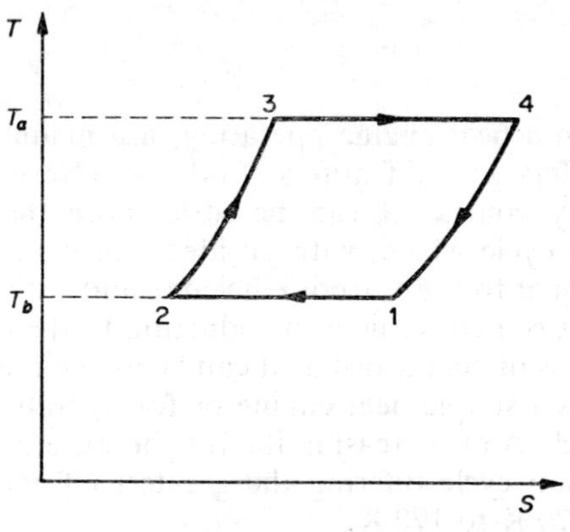

FIGURE 18.2

q_{41} were to be stored and used to supply the heat input q_{23} later in the cycle, the net heat transfer during the whole cycle would occur only during the isothermal processes. In this case equation (18.1) reduces to

$$\eta = \frac{w_{34} - w_{12}}{q_{34}} \tag{18.2}$$

Therefore , for an ideal gas

$$\eta = \frac{RT_a \ln (V_4/V_3) - RT_b \ln (V_1/V_2)}{RT_a \ln (V_4/V_3)}$$

Since $V_4 = V_1$ and $V_3 = V_2$

$$\eta = \frac{T_a - T_b}{T_a}$$

This is the Carnot efficiency and moreover in this case is obtained in a cycle with a very much larger enclosed area on the p–v diagram than a Carnot cycle offers. The Stirling cycle, then, is capable in principle both of the maximum possible cycle efficiency for an engine operating between two fixed temperatures and also of yielding a work output per cycle which is comparable with that obtained by the use of the Otto or Diesel cycles, neither of which is capable, even in ideal circumstances, of an efficiency as high as the Carnot cycle. The difficulty is that of achieving in practice an efficient storage of the heat rejected q_{41} for re-use as the heat injected q_{23}. This process is attempted in a *regenerator* consisting, usually, of a matrix of wire gauze or fine tubes made from a highly conducting metal through which the working fluid passes in each of the constant volume processes, in one case giving up heat to the matrix and in the other removing heat from the matrix. For these heat transfer processes to be efficient the working fluid must pass slowly through the matrix thus making the time taken to complete each cycle rather long. Even if the regeneration is perfect (which will not be the case) the number of cycles completed in unit time will be relatively small, so that the power delivered by this type of engine will not compare with that delivered by a less efficient but higher speed engine operating on an Otto or Diesel cycle. However, when the cycle is reversed and operated as a refrigerator these unfavourable comparisons no longer apply and modern Stirling cycle refrigerators are the best available for producing temperatures in the range 70 K to 170 K both from the point of view of efficiency judged as the amount of work needed to produce a given mass of, for example, liquid nitrogen, and in the time taken for this process. In addition modern refrigerators working on the Stirling cycle are now capable of a sufficiently high degree of reliability to be feasible for sustained operation in commercial applications.

Experimental Details

Test Equipment. At least one simple version of a single cylinder Stirling cycle engine designed specially for teaching purposes is available commercially. The principle of operation, which is illustrated in Figure 18.3, hinges on dividing the cylinder of the engine into two halves, one 'hot' and one 'cold', with the temperature of the walls of the cold half maintained constant by water cooling. The volume of the air enclosed in the cylinder is varied by the movement of the piston P. In addition there is a second 'piston', the displacer D, which moves up and down out of phase with P, and whose purpose is to transfer the working fluid (air) between the two halves of the cylinder without contributing to any change in the volume of the air in the cylinder, which is determined by the motion of P alone. The displacer D carries an axial regenerator matrix made of copper wool which alternately absorbs heat from the air passing through it in one part of the cycle of operations and rejects heat to the air passing through it in the next half of the cycle. The movement of the displacer D is controlled by an eccentric linkage on the drive shaft which arranges for movements of D and P to be approximately 90° out of phase at all times. In this way an approximation to a Stirling cycle is possible without the complete regenerative heat transfer which makes the efficiency of the cycle equal to that of a Carnot cycle—the appropriate equation for the cycle efficiency, assuming it to be a true Stirling cycle, is equation (18.1).

To improve the heat transfer to the cooling water the lower face of the displacer D is water cooled through pipes passing axially along the connecting rod between it and the driving shaft. Into the upper half of the cylinder it is possible to fit either an electric heating coil A (when the apparatus is to be used as a heat engine) or a thermometer (to investigate the use of the apparatus as a refrigerator or a heat pump). For use as a heat pump the relative

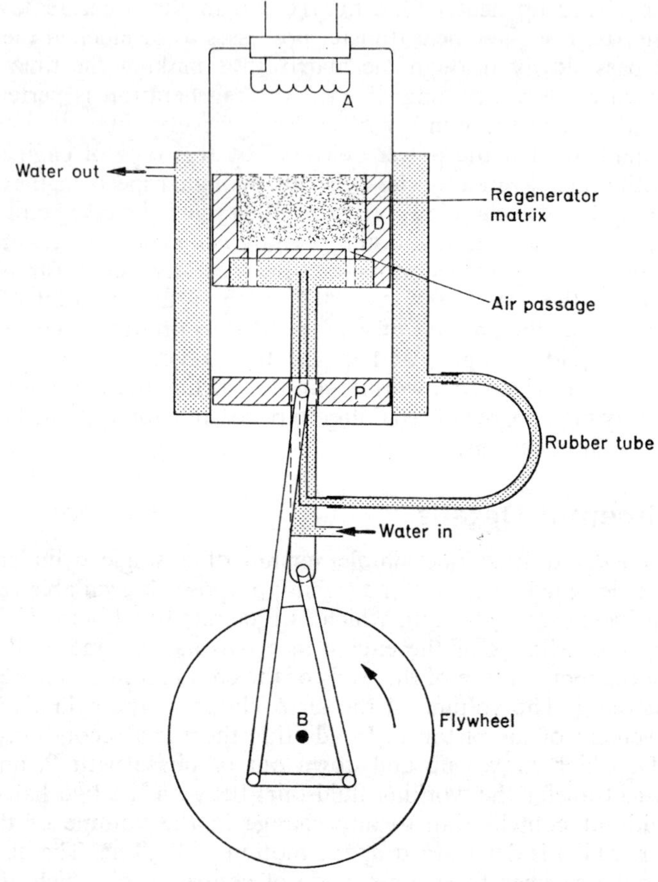

FIGURE 18.3

phase of the motion of P and D must be reversed by reversing the direction of rotation of the driving shaft.

A pressure tapping point provided via an axial hole in the rod connecting the piston P to the drive shaft makes it possible to plot a p–v diagram for the operation of the apparatus.

Procedure. To use the apparatus as a heat engine, the heating coil A should be fitted in the upper end of the cylinder. Starting with the displacer D at top dead centre most of the air inside the cylinder is in the lower half which is maintained at a lower temperature than the upper half by the water-cooled jacket. As the flywheel rotates about its axis B, the displacer D moves downwards and the piston P upwards compressing the air in the cylinder and transferring that in the lower half through the regenerator matrix to the upper half of the cylinder, some of the heat of compression being transferred to the cooling water and some (in the first stroke only) to the regenerator matrix. This corresponds approximately to isothermal compression of the gas. When the air enters the upper half of the cylinder it is heated by coil A and its pressure increases, this part of the cycle corresponding roughly to increasing the pressure at constant volume. Clearly the heat input at this stage does not come from the regenerator and there is no question of a Carnot efficiency being approached. The regenerator in this case serves only to improve the use made of the heat available; its action is most important in the use of the apparatus as a heat pump and as a refrigerator.

As the pressure of the air in the cylinder increases it acts on the face of the piston P forcing it downwards in the working stroke. At the same time the regenerator matrix is heated by the hot air being transferred from the upper to the lower half of the cylinder where its temperature falls as it loses more heat to the cooling water. This corresponds to the expansion stroke 3–4 and the constant volume part of the cycle 4–1 in Figure 18.1. The continued supply of heat to the coil A during the expansion stroke tends to counteract the cooling which would result from the expansion of the gas helping to make the first part of the stroke approximately isothermal. The heat transferred to the regenerator matrix and the cooling water in the lower half of the cylinder corresponds to the part of the true Stirling cycle in which heat is rejected at constant volume.

This cycle of operations then repeats itself; this time, and for all subsequent strokes, the regenerator matrix is hot and during the downward movement of the displacer gives out heat to the air being transferred from the lower to the upper half of the cylinder. It thus acts as a preheater before the main heat input from the coil A.

The engine should execute about five complete cycles in one second, giving a power output of some 10 watts for an electrical input of about 200 watts. The power output can be measured with a small brake drum attached to the axis of the flywheel where the torque will be about 0·25 N m. The electrical power input can easily be measured, from the voltage applied to the coil and the current flowing. Hence the efficiency of the cycle can be calculated.

To use the apparatus as a refrigerator the heating coil A should be replaced by a test tube containing some water whose temperature is measured with a suitable thermometer. The flywheel should be rotated in the same direction as before using a small electric motor whose power consumption can be measured. The cycle of operations is the same as before with the difference that after the initial compression of the gas and the transfer to the cooling water and to the regenerator of some of the heat of compression, the air expands in the upper half of the cylinder and, in consequence, cools and removes some of the heat from the water in the test tube. As the displacer rises, the cold air passes through the regenerator from which it removes most of the stored heat of compression and on entering the lower half of the cylinder is restored to its initial temperature by absorbing heat from the cooling water. The cycle of operations is then repeated and in this way there is an effective continual withdrawal of heat from the upper half of the cylinder. This heat is in fact transferred to the water which cools the lower half of the cylinder as the following argument indicates.

At the end of each thermodynamic cycle undergone by the air in the cylinder it must be returned to its initial state if heat losses, friction, etc., are neglected. The net heat transfer to the air must then be zero. Let the total heat flow from the air in one cycle be Q_1 to the cooling water and Q_R to the regenerator; let the total heat absorbed be Q from the body to be cooled (e.g. a test tube of water in the upper half of the cylinder), Q_R from the regenerator (assumed perfect) and Q_2 from the water. Then

$$-Q_1 - Q_R + Q + Q_R + Q_2 = 0$$

Hence

$$Q = Q_1 - Q_2$$

which is the net heat flow into the cooling water.

In all this the regenerator plays an important part, since in it the heat transfer is good. On leaving it the air temperature should be very close to its temperature before compression. In the subsequent expansion the temperature reached in the top half of the cylinder will be lower than it would be without the regenerator, and so the heat transferred from the body being cooled is as large as possible.

Finally, by reversing the direction of rotation the apparatus can be made to act as a heat pump. In this case when the displacer D is at top dead centre the air expands in the lower water cooled half of the cylinder, its temperature falls and it absorbs heat from the water. The air is then transferred to the upper half of the cylinder, cooling the regenerator *en route*, and compressed. In this part of the cycle, it can deliver heat to any body inserted in the top half of the cylinder. When the air is returned to the lower half of the cylinder it transfers heat to the cool regenerator matrix. The air temperature is then low enough at the beginning of the expansion stroke of the cycle for its temperature, in the absence of the water jacket, to reach a temperature lower than that at the beginning of the cycle. The initial thermodynamic state is restored by absorption of heat from the water flowing in the jacket. This heat together with some of that stored in the regenerator matrix is available for release in the upper half of the cylinder. In this way there is a steady transfer of heat from the water in the jacket to the top half of the cylinder. It is easy to boil water in a test tube inserted into the top half of the cylinder.

Further Experiment

Using the pressure tapping on the main cylinder connecting rod together with a pressure transducer and a suitable *x*-axis sweep proportional to the movement of the main piston P, a p–v diagram could be plotted and the enclosed area used to estimate work done by the apparatus when acting as a heat engine.

References

1. Kohler, 'The Stirling Refrigeration Cycle', *Scientific American*, April 1965.
2. Rogers, G. F. C. and Mayhew, Y. R. *Engineering Thermodynamics, Work and Heat Transfer*, Longmans, 1970.

19

The Single-stage Vapour Compression Refrigerator

For the present experiment a vapour compression refrigerator was chosen as a readily available commercial unit which could easily be modified suitably. The various components are separated and joined by pipes fitted with appropriate instruments for measuring temperature, pressure and flow rate; in place of the usual 'cold space' a calorimeter is provided to make possible measurements of the quantity of heat removed by the refrigerator; finally a water-cooled condenser is used so that the heat rejected from the refrigerator circuit can also be measured.

Theory

Practical refrigerator cycles are based on a reversed vapour power cycle in which the pressure reduction which could be obtained in an isentropic expansion is achieved more simply by using instead a throttling valve giving a decrease in pressure in an irreversible isenthalpic process. The basic cycle is composed of four flow processes illustrated in Figures 19.1, 19.2, and 19.3, on *p–v*, *T–s* and *p–h* diagrams respectively. Each process takes place in a separate unit, and the four units being connected together as shown in Figure 19.4. Heat is transferred from the cold space to the working fluid in the evaporator, from which the working fluid emerges either as a saturated vapour at a temperature T_a or as a slightly superheated vapour. It is then compressed approximately isentropically, its temperature rising to T_c, after which it enters the condenser where it is cooled in a constant pressure flow process and leaves as a saturated liquid at a temperature T_3. The working fluid then passes through the throttling valve and it becomes a mixture of saturated liquid and vapour at a temperature T_a. It then re-enters the evaporator and the process is repeated.

The design of a refrigerator cycle is normally such that changes of kinetic and potential energy are negligible in relation to the quantities of heat transferred so that the steady flow energy equation in the usual notation reduces to

$$h_1 - h_2 = w - q \qquad (19.1)$$

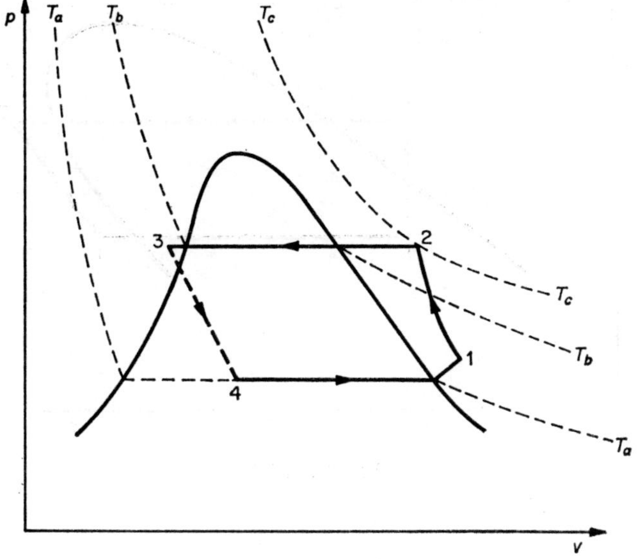

FIGURE 19.1

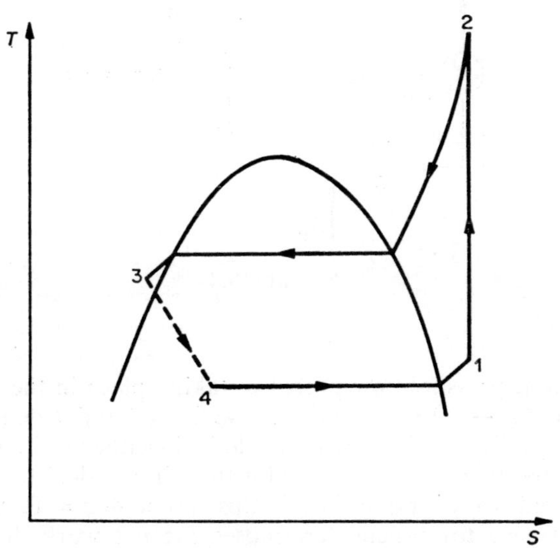

FIGURE 19.2

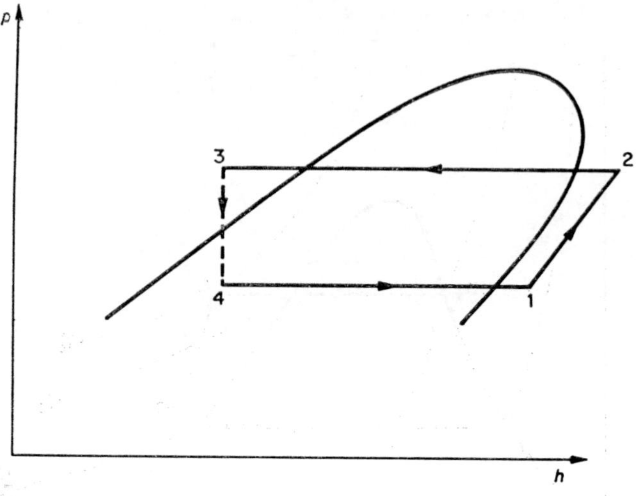

FIGURE 19.3

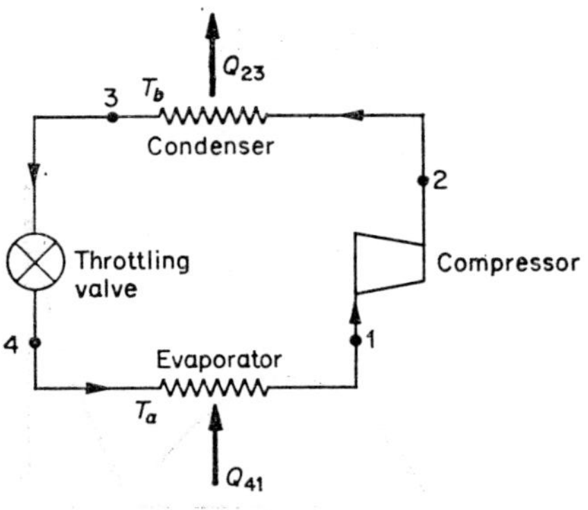

FIGURE 19.4

In the constant pressure flow processes taking place in the condenser and the evaporator $(h_i - h_f) = q$, where the suffixes i and f denote initial and final states respectively of the working fluid in either unit. Hence in both cases the external work done is zero. Further, in an ideal throttling process again no external work is done since in this case $w - q = 0$, and the process is adiabatic. Hence for practical purposes the net work done during the whole cycle is that done by the compressor. Since the compression is approximately isentropic, the net work done can be obtained from equation (19.1) with $q = 0$.

The coefficient of performance c of the cycle is defined as the ratio of the quantity of heat removed from the cold space to the net work done in the cycle, i.e.

$$c = \frac{Q_{41}}{W_{\text{net}}} = \frac{h_1 - h_4}{h_2 - h_1}$$

Since 3 → 4 is an isenthalpic process,

$$c = \frac{h_1 - h_3}{h_2 - h_1}$$

All these quantities can be found from tables of properties of the working fluid provided the temperatures are known at points 2 and 3 and both temperature and pressure at point 1.

Experimental Details

Test Equipment. The components of a commercial vapour compression refrigerator have been separated and re-connected as shown in Figure 19.6,

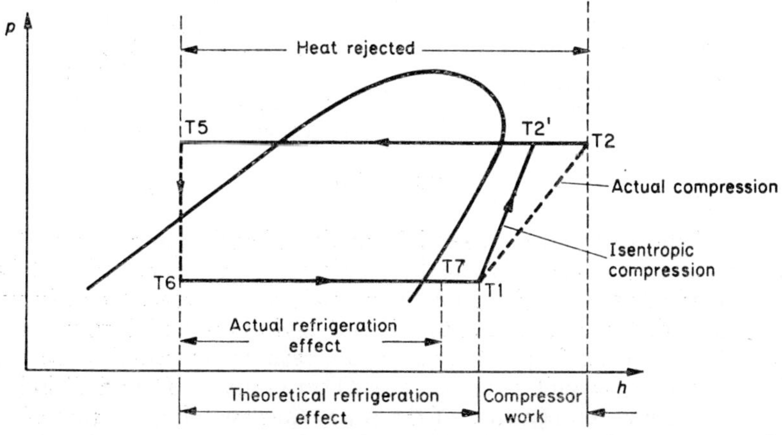

FIGURE 19.5

with temperature and pressure measuring points (denoted respectively by T1 to T9 and by P1 and P2) fitted into the connecting pipes. In addition a 'calorimeter' and a water-cooled condenser are needed. The pressure and temperature inside the calorimeter are measured and in the condenser the heat absorbed is measured from the flow rate of the water and the temperature difference between the water entering and leaving the condenser.

The calorimeter is a sealed insulated vessel in which are placed the evaporator of the refrigerator circuit and two electrical heating coils capable of supplying heat to the calorimeter at two discrete rates chosen to match the capacity of the refrigerator used. The whole calorimeter is filled with liquid

Freon which acts as the medium from which heat is abstracted in the refrigerator. If the temperature of the calorimeter is maintained at an effectively constant value when one or both of the electrical heating coils and the refrigerator are in operation then the rate of heat removal from the calorimeter by the refrigerator must be equal to that at which heat is supplied by the electrical heaters.

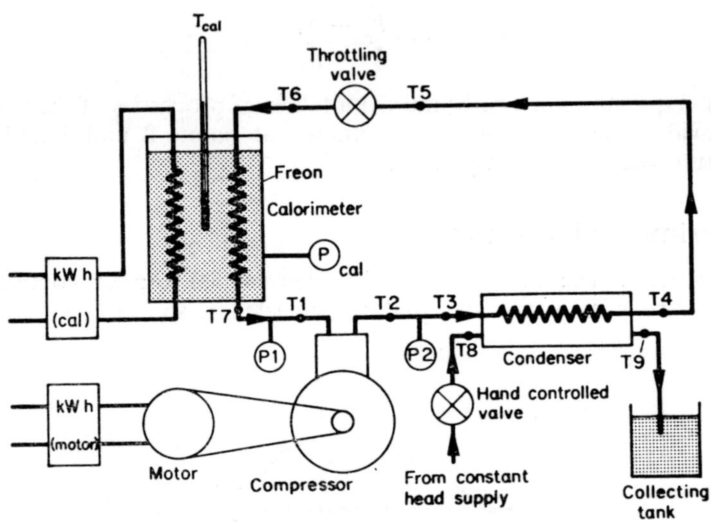

FIGURE 19.6

Instead of the standard commercial practice of exchanging heat with the atmosphere the condenser coils of the refrigerator in this experiment have been mounted in a separate insulated water-cooled condenser. A steady water supply is provided from a constant head tank and the rate of flow, which can be controlled by a hand-operated valve, is measured by weighing the water passing through the circuit in a given time. Thermometers T8 and T9 indicate the inlet and outlet temperature of the water thus enabling the rate of heat rejection to the cooling water in the condenser to be measured directly.

Two watt meters, or kW h meters, are needed for measuring directly the power supplied to the electric heaters and to the compressor motor. The latter, as has already been noted, provides virtually all the work done during the refrigerator cycle.

Procedure. First open fully the condenser cooling-water control valve, fill up the constant head supply tank and run water through the condenser at the maximum possible flow rate. Switch on the refrigerator compressor and let the refrigerator run until the temperature of the Freon in the calorimeter has fallen to about −7 °C, when its pressure will be about 2·5 bars. Then switch on one of the calorimeter heaters and by means of the hand-operated control valve reduce the flow of cooling water to the condenser until T_b, the temperature at which condensation occurs, has risen to about 20 °C; then the saturation temperature of Freon is about 5·5 bars. Since in an ideal cycle the pressure at the exit from the compressor is the same as the saturation pressure at the condensation temperature, reducing the flow of cooling water to the condenser has the effect of increasing the compressor delivery pressure. By adjusting the flow of condenser cooling water, keep the compressor delivery pressure, as indicated by P2, constant at about 5 bars, and after about 20 minutes when conditions are reasonably steady, carry out a 30 to 40 minute run, reading at 5 or 10 minute intervals pressures P1 and P2 and temperatures T1, T2, T5, T6 and T7. Also read the kW h meters at the beginning and end of the measurement.

The whole set of readings can then be repeated with both the electric heaters in the calorimeter switched on.

For each set of readings the actual overall coefficient of performance of the refrigerator is given by

$$\text{coefficient of performance} = \frac{\text{kW h supplied to heaters in calorimeter}}{\text{kW h supplied to compressor motor}}$$

If the refrigerator cycle is drawn on a p–h diagram as shown in Figure 19.5, it is easy to find the following coefficients, using tables of values for the enthalpy of the refrigerator working fluid:

$$\text{diagram coefficient of performance} = \frac{h(T7) - h(T6)}{h(T2) - h(T1)}$$

$$\text{theoretical coefficient of performance} = \frac{h(T1) - h(T6)}{h(T2') - h(T1)}$$

Also the dryness fraction of the working fluid after throttling, when its temperature is $T6$, can be determined by assuming the throttling to be ideal.

Further Experiment

A useful extension is to include in the circuit a device to indicate directly the rate of flow of the refrigerator working fluid. This can then be compared directly with values for the flow rate of the refrigerant calculated from temperatures $T6$ and $T7$ and the heat input to the calorimeter and also from temperatures $T3$, $T4$, $T8$ and $T9$, and the flow rate of the condenser water. Finally a full heat balance can be made for the whole circuit.

20

A Performance Test for an Internal Combustion Engine

No undergraduate thermodynamics course would be complete without an experiment whose aim is to investigate some aspect of the performance of an internal combustion engine. It is difficult to devise a single experiment which is likely to be useful in all laboratories, since so much depends on the apparatus available. Where variable compression engines are available an investigation of the effect on thermal and volumetric efficiency of changing the compression ratio, but with the swept volume necessarily remaining constant, offers quite an attractive experiment. With a perfectly standard engine the choice of experiment is virtually restricted to measurements of design parameters and to measurements of efficiency, both of which should be within the capability of most laboratories. The experiment described is a straightforward investigation of an internal combustion engine (which could use either spark ignition or compression ignition), many variations of which will be possible.

Theory

(*a*) *Thermal Efficiency*. In all internal combustion engines the basic definition of thermal efficiency η_{th} is

$$\eta_{th} = \frac{\text{work output in unit time}}{\text{heat input in unit time}} = \frac{\dot{W}}{\dot{Q}} \tag{20.1}$$

where the heat input is derived from the combustion of fuel which releases an amount of heat Q_c for each mass unit of the fuel which undergoes complete combustion. Thus

$$\dot{Q} = \dot{M}_f Q_c \tag{20.2}$$

where $\dot{M}_f$ is the rate at which fuel is supplied to the engine, which can easily be measured directly.

The work output in unit time can be measured either from an indicator diagram which gives the 'indicated power', or directly from the torque applied to the shaft of the engine, i.e. the brake power, as measured when the engine is turning at constant speed against a known braking torque. Taking the work output to be the brake power (B.P.) and putting $\dot{W} = \text{B.P.}$ it follows that

$$\eta_{\text{th(brake)}} = \frac{\text{B.P.}}{\dot{M}_f Q_c} \tag{20.3}$$

This can equally well be written in terms of the rate at which air is supplied to the engine $\dot{M}_a$, which is again easily measured, and F, the fuel/air mass ratio, so that

$$\dot{M}_f = \dot{M}_a F \tag{20.4}$$

and hence

$$\eta_{\text{th(brake)}} = \frac{\text{B.P.}}{\dot{M}_a F Q_c} \tag{20.5}$$

(*b*) *Specific fuel consumption.* The rate at which an engine consumes fuel can, with advantage, be written in terms of the fuel consumed per unit power output, i.e. the specific fuel consumption (which may refer either to brake or indicated power). Thus

$$\text{brake specific fuel consumption} = \frac{\dot{M}_f}{\text{B.P.}}$$

$$= \frac{1}{\eta_{\text{th(brake)}} Q_c} \tag{20.6}$$

(*c*) *Volumetric efficiency and air mass ratio.* The volumetric efficiency V_E of an engine is defined by the equation

$$V_E = \frac{\text{volume of air drawn in per stroke}}{\text{volume swept by piston per stroke}} \tag{20.7}$$

Since space has to be provided in the cylinder to allow the inlet and exhaust valves to open, the total cylinder volume is greater than the swept volume. It follows that the residual gases compressed into this additional volume at the end of one cycle must expand until the pressure inside the cylinder is less than the inlet pressure before a fresh supply of fuel/air mixture in spark ignition (S.I.) engines, or of air in compression ignition (C.I.) engines can be admitted when the inlet valve opens. Thus V_E will depend on the cylinder dimensions, the compression ratio of the engine, which determines the pressure of the residual gases, and on the inlet pressure chosen.

The actual numerical values used in equation (20.7) will clearly depend on the conditions chosen to evaluate the volume appearing in the numerator. If this volume is evaluated at s.t.p. under which conditions the density of air is ρ then

$$V_E = \frac{\rho\ (\text{volume air drawn in per stroke corrected to s.t.p.})}{\rho V_s}$$

where V_s denotes the swept volume. Thus

$$V_E = \frac{\text{mass air drawn in per stroke}}{\text{mass needed to fill } V_s \text{ at s.t.p.}}$$

$$= \sigma$$

where σ is called the air mass ratio. Since

$$\sigma = \frac{\text{mass air drawn in per stroke}}{\rho V_s}$$

it follows that

$\rho\sigma$ = mass of air drawn into unit volume of the swept volume V_s per stroke

If, now, Z denotes the fraction of the air present in unit volume of the swept volume which is actually used for combustion, and M is the mass of air needed for complete combustion of unit mass of fuel releasing an amount Q_c of heat, then

$\frac{Z\rho\sigma}{M} Q_c$ = heat energy released in unit volume of the swept volume

Hence the specific output of the engine, measured in terms of brake power, is equal to the shaft work done per unit volume of the swept volume. Thus

$$\text{specific output} = \frac{Z\rho\sigma Q_c}{M}\eta_{\text{th(brake)}}$$

$$= \frac{\text{b.m.e.p.} V_s}{V_s}$$

$$= \text{b.m.e.p.}$$

where b.m.e.p. is the mean effective pressure during one cycle, measured from the brake power output. Hence

$$\text{b.m.e.p.} = Z\left(\frac{Q_c}{M}\right)\rho\sigma\eta_{\text{th(brake)}} \qquad (20.8)$$

and it follows since Q_c/M and ρ are constants that the specific output of the engine is proportional to $Z\sigma\eta_{\text{th(brake)}}$.

For compression ignition engines Z never exceeds 0·65, while for spark ignition engines it never exceeds 0·83; a typical value of σ for both C.I. and S.I. engines is 0·8; typical values for $\eta_{\text{th(brake)}}$ are 25 to 35 per cent for S.I. engines and 30 to 45 per cent for C.I. engines.

Experimental Details

Test Equipment. In addition to the standard petrol or diesel engine to be tested the following ancillary apparatus is needed:

(*a*) A dynamometer to measure the torque developed by the engine.

(*b*) A facility similar to that illustrated in Figure 20.1 for measuring the rate at which the engine uses fuel. This can be accomplished by measuring with a stop-watch the time taken for a fixed volume of fluid, between two marks on a pipette, to pass on to the engine fuel pump.

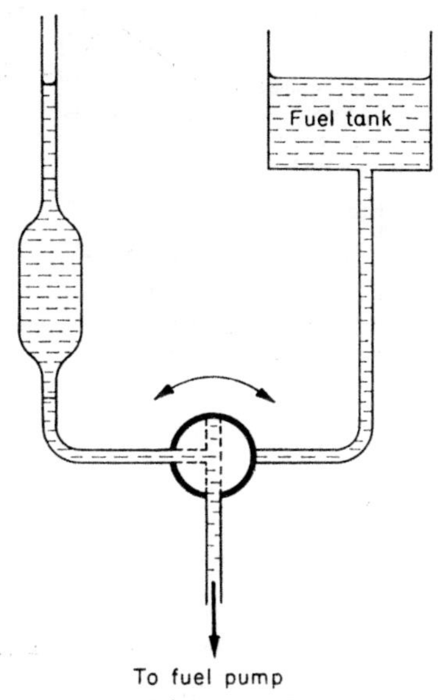

FIGURE 20.1

(*c*) A facility similar to that shown in Figure 20.2 for supplying the cooling system of the engine with a constant flow of water whose inlet temperature is effectively constant. Oil-filled thermometer pockets suitable for either a differential thermocouple pair or for two separate mercury-in-glass thermometers should be fitted to measure inlet and outlet water temperatures. In addition a pivoted funnel is needed to enable the water leaving the engine to be directed for a measured time into a collecting vessel, which is weighed before and after to determine the water-flow rate.

(*d*) A facility to measure the rate at which air is consumed by the engine is needed. This is conveniently done with a plenum chamber fitted using either a venturi meter or an orifice plate in the air-intake line, measuring the pressure drop across this flow constriction and also the values of the temperature and absolute pressure at the venturi, or orifice. The venturi, or orifice plate, will of course need to be calibrated prior to the experiment. In addition the temperature in the inlet and exhaust manifolds should be measured by mounting thermocouples inside each manifold; ideally these temperatures should be read simultaneously at several points in each manifold and a mean taken.

(*e*) Facilities will be required for taking an indicator diagram for one of the cylinders. Various ways of doing this are available; almost any standard method would be suitable for this experiment.

(*f*) A revolution counter, tachometer-generator, or other device to measure the speed of rotation of the engine shaft must be available.

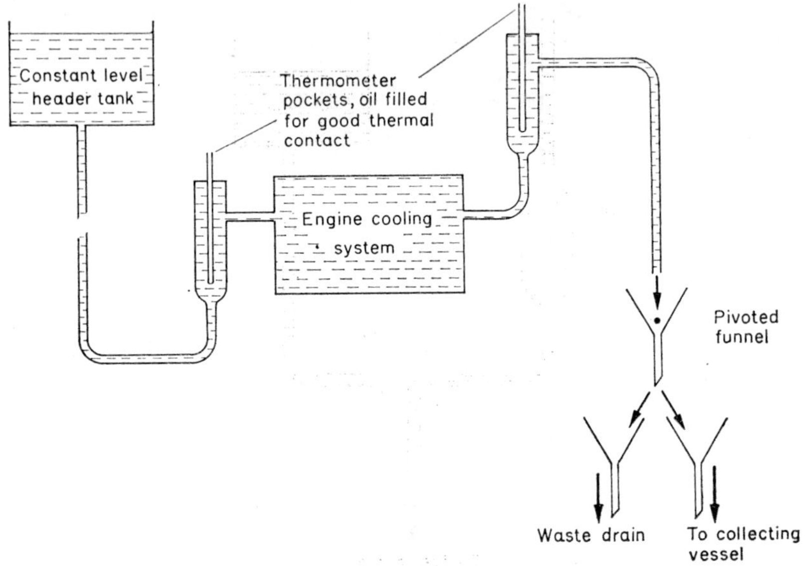

FIGURE 20.2

Procedure. Set the engine running and with a known torque applied to the shaft allow it to run until conditions are steady with the engine running at its design speed, which will be kept constant during the experiment. Measure the rate of fuel consumption $\dot{M}_f$ by finding the time taken for the level in the pipette shown in Figure 20.1 to fall from the upper to the lower fixed mark. Measure the pressure drop across the venturi (or orifice) in the air supply line to give $\dot{M}_a$. Also, if the calculation of a heat balance is to be included, measure the inlet and outlet temperatures of the cooling water and the temperatures in the inlet and outlet manifolds of the engine. Then change the load applied to the engine and adjust its speed until it returns to its initial value. After conditions have become steady repeat the above measurements. This should be done for about six different loads.

In each case make the following calculations:

(i) Determine $\eta_{\mathrm{th(brake)}}$ using equation (20.3) taking an appropriate valuc of Q_c from tables and calculating the brake power from the torque on the dynamometer.

(ii) Using equation (20.4) and the measured values of $\dot{M}_f$ and $\dot{M}_a$, find the fuel/air mass ratio F.

(iii) Find the brake specific fuel consumption directly from the measured values of $\dot{M}_f$ and the brake power.

(iv) Calculate the value of b.m.e.p. using known values of the swept volume V_s of the cylinder and the measured values of the output of the engine and the rate of rotation of the crankshaft. Compare this with the value obtained from the indicator diagram and hence find the mechanical efficiency of the engine.

(v) Find the value of σ from the measured value of $\dot{M}_a$ and the number of revolutions N per second executed by the crankshaft when the engine is running steadily, i.e.

$$\sigma = \frac{2\dot{M}_a}{N} \cdot \frac{1}{\rho V_s}$$

(vi) Use equation (20.8) to find the value of Z.

This part of the experiment enables graphs such as those shown in Figures 20.3 and 20.4 to be plotted showing the way in which brake specific fuel consumption and engine efficiency vary with brake power at a given speed. It also gives typical values of F and Z when the engine is running under normal adjustment.

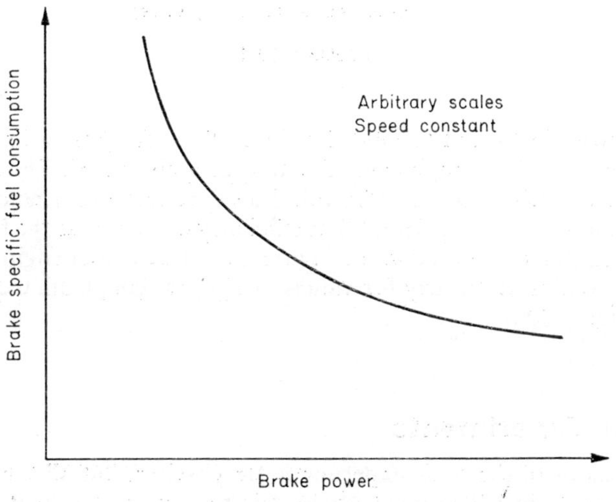

FIGURE 20.3

The effect on efficiency of varying F can be explored by setting the engine running at a constant speed against a constant load and deliberately varying F whilst keeping constant the rate of rotation of the crankshaft. It is fairly simple to vary F in the case of a Diesel engine since this is the normal method of power control. For a petrol engine without fuel injection it is less easy, but using the carburettor mixture control should give sufficient variation for this experiment. For each chosen value of F, the brake power, $\dot{M}_f$ and $\dot{M}_a$ should be measured. Equation (20.3) then gives $\eta_{th(brake)}$ and equation (20.4) the value

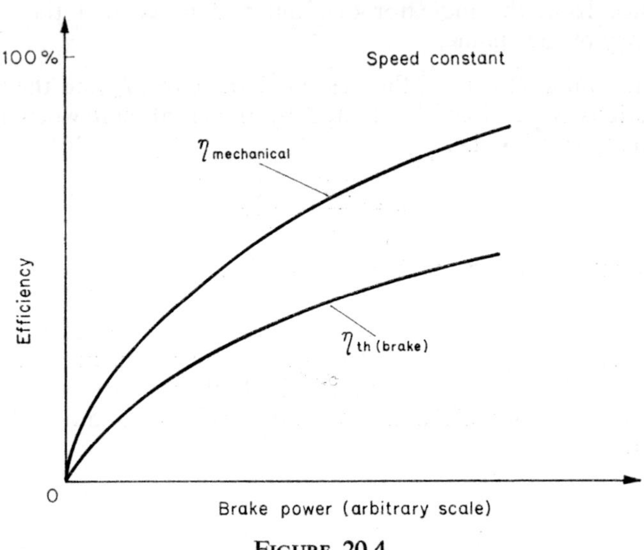

FIGURE 20.4

of F. It is possible to find tabulated values of the fuel/air ratio F_c which is chemically correct for complete combustion of a given fuel. Then a graph of brake thermal efficiency against F/F_c will show the effect on efficiency of varying F. For S.I. engines it will be found that efficiency decreases as the fuel/air ratio increases, qualitatively as shown in Figure 20.5. For Diesel engines the efficiency decreases almost linearly for values of F/F_c ranging from 0·2 to unity as shown in Figure 20.6.

Further Experiments

Many extensions of the basic experiment are possible, but all are liable to be very time consuming. The most obvious is to construct a heat balance for the engine which should show a marked increase in the heat unaccounted for as the output of the engine increases.

Another possibility for fuel injection engines is to throttle the inlet manifold until the inlet valves no longer admit sufficient air for complete combustion. The situation is then similar to that occurring in high speed engines if the inlet valve apertures become 'choked' at the higher engine speeds. The effect on V_E or σ of this restricted air supply will be of interest.

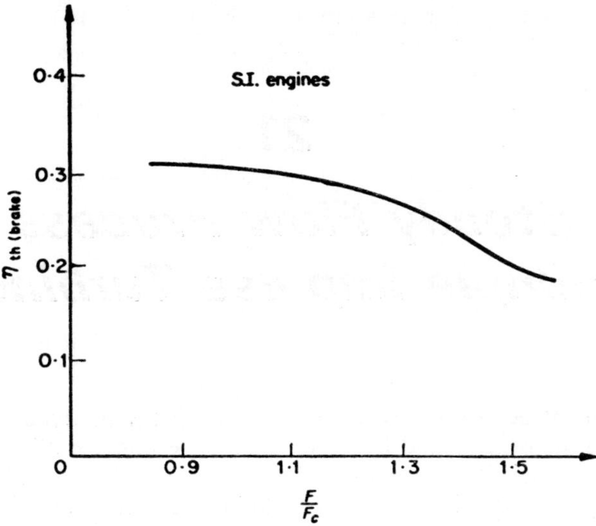

FIGURE 20.5

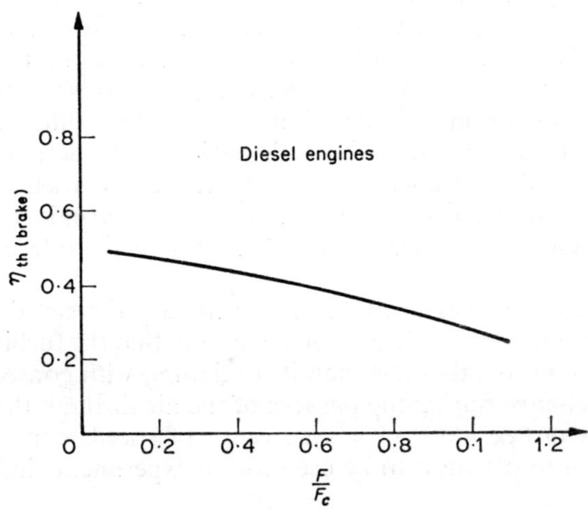

FIGURE 20.6

References

1. Rogers, G. F. C. and Mayhew, Y. R. *Engineering Thermodynamics: Work and Heat Transfer*, Longmans, 1970.
2. Pye, D. R. *The Internal Combustion Engine*, Oxford University Press, 1937.
3. Taylor, C. F. *The Internal Combustion Engine in Theory and Practice*, vol. 1, John Wiley, 1960.

21

The Steady Flow Process in a Simple Impulse Turbine

This experiment demonstrates the application of thermodynamic arguments to a steady flow process and is based on an experiment outlined by Benson, Horlock and Ryley [1]. The particular flow process described is the passage of air through a simple impulse turbine and requires little auxiliary apparatus other than a suction fan, some ducting, a band brake and apparatus to measure pressure and temperature. The latter may present some difficulty as the temperatures of the air at the inlet and outlet to the turbine are needed. The maximum difference of these two temperatures is likely to be only of the order of 1 °C, so temperature measurements must be made with some care if the error in the difference of these temperatures is not to be excessive. Some control over the magnitude of this temperature difference is possible by adjustment of the combination of turbine and suction fan which is chosen. The fan, for instance, could vary from one used on a domestic vacuum cleaner to a wind tunnel fan; indeed one convenient way of setting up the experiment is by attaching the apparatus to the inlet side of a small wind tunnel.

The experiment involves using the brake to vary the speed of rotation of the turbine over a range sufficiently wide to ensure that the turbine is operating at an efficiency considerably less than its optimum, with consequent increase in the entropy change during the passage of the air through the turbine. Any turbine which can operate over a wide range of speeds can, in principle at least, be adapted to permit it to be used for an experiment similar to the one described.

Theory

The steady flow energy equation applied to unit mass of fluid flowing through any arbitrary bounded system can be written in the form

$$\Delta h + \Delta(\text{K.E.}) + \Delta(\text{P.E.}) = w - q \tag{21.1}$$

In this equation Δh, Δ(K.E.) and Δ(P.E.) denote respectively the differences between the values at the inlet and outlet of the system of the specific enthalpy of the flowing fluid, the kinetic energy and potential energy of unit mass of the fluid and w and q are respectively the net work and heat transfer to unit mass of the flowing fluid in unit time. For turbines there is usually no change in the potential energy between inlet and outlet. Hence, using suffixes i and f to denote initial and final values of any fluid property, and using C to denote velocity, equation (21.1) becomes

$$h_i - h_f + \tfrac{1}{2}(C_i^2 - C_f^2) = w - q \qquad (21.2)$$

Each stage of a turbine is composed of a set of fixed nozzles which accelerate the flowing fluid and direct it on to a set of rotating blades mounted on a disc as illustrated in Figures 21.1, 21.2 and 21.3. Figure 21.2 shows a section through the nozzles and blades at the mean radius and where the linear tangential velocity of the blades in the plane of rotation is U (i.e. U = angular velocity $\omega \times$ mean radius r).

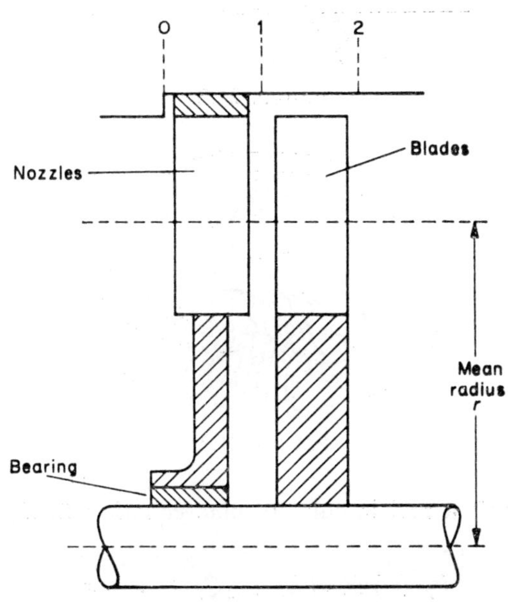

FIGURE 21.1

For any turbine its degree of reaction λ is defined by

$$\lambda = \frac{\text{enthalpy drop in rotor}}{\text{enthalpy drop in stage}}$$

In an ideal impulse turbine $\lambda = 0$ so that all the enthalpy drop occurs in the nozzles. In practice, however, λ has a small value for an impulse turbine and there is an enthalpy drop and a pressure drop both in the nozzles and in the blades, though the latter is smaller. This is illustrated by the line AB in Figure 21.7.

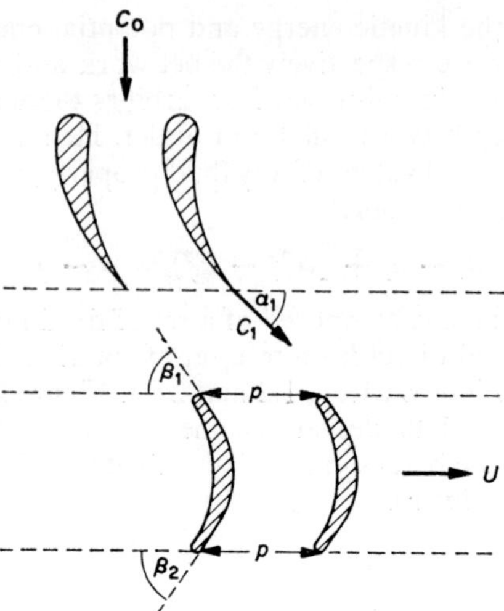

FIGURE 21.2

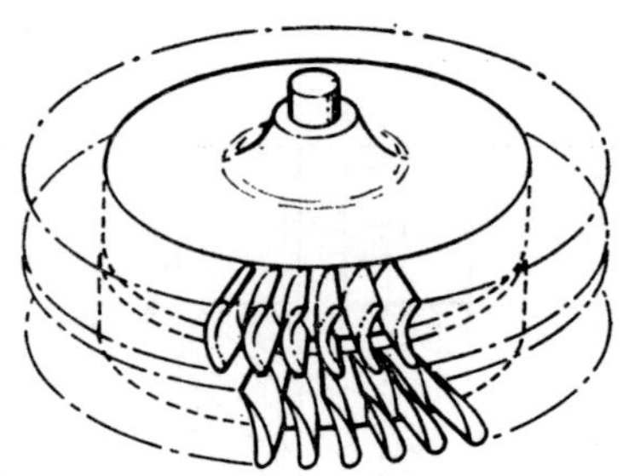

FIGURE 21.3

For flow through a real impulse turbine equation (21.2) applied to the nozzles gives

$$h_0 + \tfrac{1}{2}C_0^2 = h_1 + \tfrac{1}{2}C_1^2 + w - q \tag{21.3}$$

Since the nozzles do not move, no external work is done; further, if the flow is assumed to be sufficiently rapid for there to be no effective transfer of heat to or from the fluid, equation (21.3) becomes

$$h_0 - h_1 = \tfrac{1}{2}(C_1^2 - C_0^2) \tag{21.4}$$

In considering flow through the blades, where relative to a coordinate system stationary in the laboratory all the external work is done, it is very useful to apply equation (21.2) relative to a set of coordinates rotating about the turbine axis with the same uniform angular velocity ω as the rotor. The accelerations implicit when referring coordinates to rotating axes will influence the values of h_1 and h_2 equally so that any changes in these values cancel one another. Further relative to these axes the blades are stationary and no external work is done. Hence, neglecting heat transfer as before, equation (21.4) becomes

$$h_1 - h_2 = \tfrac{1}{2}(V_2^2 - V_1^2) \tag{21.5}$$

where V_1 and V_2 denote the fluid velocities entering and leaving the blades measured relative to the blades. Adding equation (21.4) and (21.5)

$$h_0 - h_2 = \tfrac{1}{2}(C_1^2 - C_0^2) + \tfrac{1}{2}(V_2^2 - V_1^2) \tag{21.6}$$

Applying equation (21.2) to the complete stage of the turbine gives

$$h_0 - h_2 = \tfrac{1}{2}(C_2^2 - C_0^2) + w \tag{21.7}$$

It follows that

$$w = \tfrac{1}{2}(C_1^2 - C_2^2) + \tfrac{1}{2}(V_2^2 - V_1^2) \tag{21.8}$$

or for a mass flow m

$$W = \tfrac{1}{2}m[(C_1^2 - C_2^2) + (V_2^2 - V_1^2)] \tag{21.9}$$

The vector diagram Figure 21.4 shows these air velocities, at the inlet and outlet of the rotor, measured at the mean radius of the turbine when its angular velocity is ω and is referred to both stationary and rotating axes. The radial components of velocity have been assumed to be negligible. The diagram is drawn in the plane containing the vectors $\mathbf{C}_1$ and $\mathbf{U}$. Vectorial subtraction of $\mathbf{U}$ from $\mathbf{C}_1$ gives $\mathbf{V}_1$, the velocity of the air relative to the blades, and this vector makes an angle β_1 with the direction of $\mathbf{U}$.

For an ideal impulse turbine $h_1 = h_2$; it follows from equation (21.5) that $V_1 = V_2$ and to permit this the inlet and outlet flow areas for the rotor are made equal by using rotor blades whose length and pitch p at inlet and outlet are identical. Then provided $\beta_1 = \beta_2$, V_1 must equal V_2. Both β_1 and β_2 are in fact measured relative to axes rotating with the blades and represent, relative to these axes, the angle between the vector $\mathbf{U}$ and the direction of the incoming air velocity vector; in other words β_1 is the value of α_1 referred to rotating axes instead of stationary axes. Since $\mathbf{U}_1$, $\mathbf{V}_1$ and $\mathbf{C}_1$ all lie in the same plane β_1 can be represented accurately as shown in Figure 21.4. Then by constructing a vector $\mathbf{V}_2$ equal in magnitude to $\mathbf{V}_1$ and making an angle β_2 $(=\beta_1)$ with the direction of the vector $\mathbf{U}$ (or, more precisely, making an angle of $(180° - \beta_2)$ with the vector $\mathbf{U}$) the magnitude of $\mathbf{C}_2$ and direction

α_2 of the velocity of the air leaving the rotor, measured relative to stationary axes, may be found.

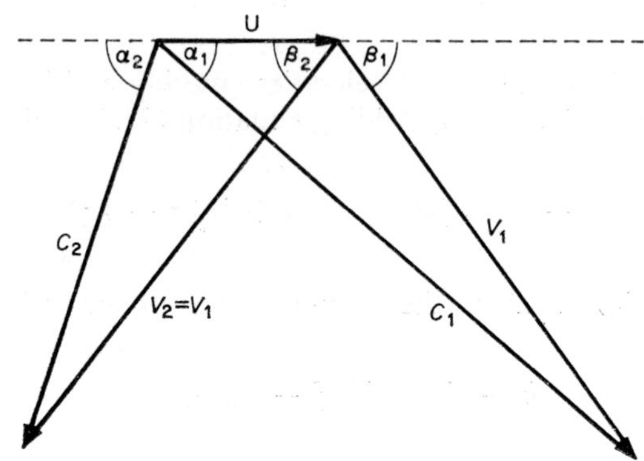

FIGURE 21.4

From Figure 21.4

$$C_1^2 = V_1^2 + U^2 + 2UV_1 \cos \beta_1$$

$$C_2^2 = V_2^2 + U^2 - 2UV_2 \cos \beta_2$$

and since $V_1 = V_2$ and $\beta_1 = \beta_2$ it follows from equation (21.9) that

$$\begin{aligned} W &= 2mUV_1 \cos \beta_1 \\ &= 2mU(C_1 \cos \alpha_1 - U) \end{aligned} \tag{21.10}$$

This equation shows that no work is done by the rotor either if

$$U = 0$$

or if

$$\cos \alpha_1 = U/C_1$$

The first condition represents the situation when the rotor is stalled; the second the situation when the rotor is 'running away', i.e. when its speed has increased sufficiently for the vectors $\mathbf{V}_1$ and $\mathbf{V}_2$ both to be perpendicular to the vector $\mathbf{U}$, making C_1 and C_2 equal.

Equation (21.10) can be used to obtain an expression for the efficiency of an ideal turbine which is clearly given by

$$\eta_{\text{ideal}} = \frac{2mU(C_1 \cos \alpha_1 - U)}{\frac{1}{2}mC_1^2} \tag{21.11}$$

$$= \frac{4U}{C_1}\left(\cos \alpha_1 - \frac{U}{C_1}\right) \tag{21.12}$$

Hence

$$\frac{d(\eta_{\text{ideal}})}{d(U/C_1)} = 4 \cos \alpha_1 - \frac{8U}{C_1}$$

$\therefore \eta_{\text{ideal}}$ is a maximum when

$$\cos \alpha_1 = 2U/C_1 \tag{21.13}$$

i.e. when the discharge velocity C_2 is axial as shown in Figure 21.5.

The value of the maximum efficiency of an ideal single stage impulse turbine follows from equations (21.12) and (21.13):

$$(\eta_{\text{ideal}})_{\text{max}} = 4(U/C_1)^2 = \cos^2 \alpha_1 \tag{21.14}$$

The magnitude of the maximum value of the work obtainable follows from equations (21.10) and (21.13) and is given by

$$W_{\text{max}} = 2mU^2 \tag{21.15}$$

In practice there will be friction between the air stream and the rotor blades so there will be small enthalpy and pressure drops in the rotor stage as well as an increase in entropy, as indicated by the line AB in Figure 21.4.

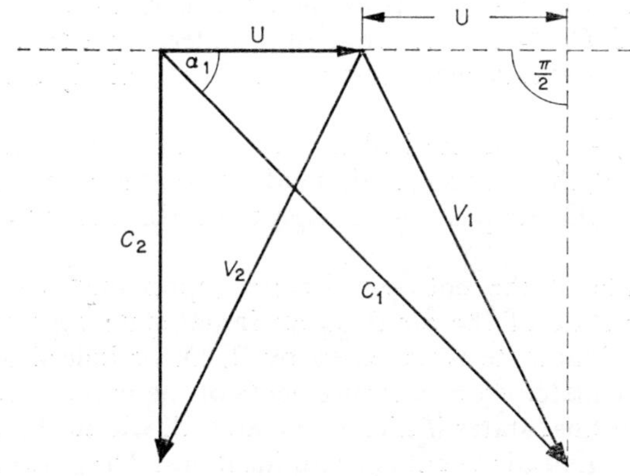

FIGURE 21.5

This change of enthalpy and pressure will increase slightly the value of $\mathbf{V}_2$ in Figure 21.4 and hence alter both $\mathbf{C}_2$ and α_2. However the main features of the behaviour of the turbine, represented by equations (21.10 to 21.15) inclusive and illustrated in Figure 21.6, will still apply, subject only to small corrections.

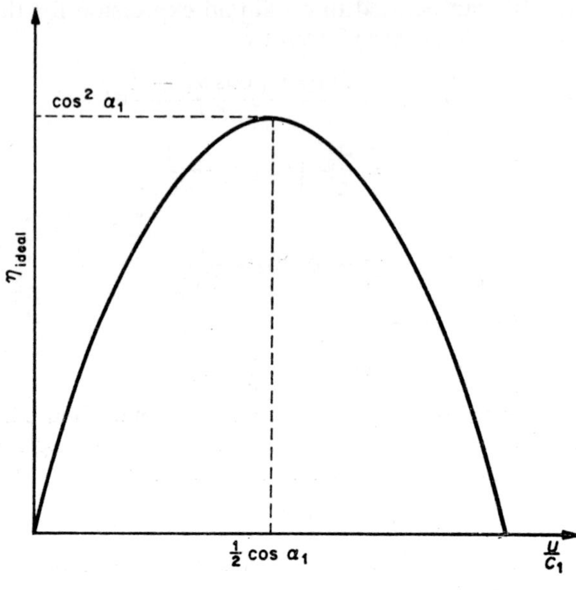

FIGURE 21.6

Returning now to Figure 21.7 where the line AB represents the actual process occurring in the turbine, points A and D represent the initial and final states of the air passing through the turbine if it is either stalled or 'running away'. In both cases, as discussed earlier, no external work is done so that provided $C_0 \simeq C_2$, the air stream undergoes a throttling process in which both enthalpy and temperature remain effectively constant while the entropy increases.

Since $(\partial h/\partial s)_p = T$ and the isobars corresponding to pressures p_0 and p_2 are nearly parallel, the points A and D indicate that both h and T are virtually unchanged when the air passing through the turbine is throttled from p_0 to p_2.

The magnitude of the entropy changes represented by changes in the thermodynamic state of the air from an initial state represented in Figure 21.4 by A to a final state represented by B, D, or indeed any other point, can easily be calculated from measurements of the pressure and temperature in the initial and final states if it can be assumed, within the limits of experimental accuracy, that air behaves as an ideal gas. The general equation

$$\begin{aligned} T\,\mathrm{d}s &= \mathrm{d}u + p\,\mathrm{d}v \\ &= \mathrm{d}h - v\,\mathrm{d}p \end{aligned}$$

applied to an ideal gas gives

$$\mathrm{d}s = c_p \frac{\mathrm{d}T}{T} - R\frac{\mathrm{d}p}{p}$$

where R is the gas constant per unit mass. Integrating from an initial state i to a final state f

$$\Delta s = c_p \ln\left(\frac{T_f}{T_i}\right) - R \ln\left(\frac{p_f}{p_i}\right) \tag{21.18}$$

If an experiment is performed in which measurements are made of static values of pressure and temperature at the inlet and outlet to the turbine (i.e. in planes 0 and 2 of Figure 21.1) it is possible to calculate:

(*a*) the enthalpy change $h_0 - h_2 = c_p(T_0 - T_2)$;
(*b*) the entropy increase $(s_2 - s_0)$ using equation (21.18).

It is also possible to vary the entropy increase by braking the rotor so that its angular velocity, and hence the value of U, is varied sufficiently to give a wide range of possible efficiencies between the two extreme conditions when the rotor is either stalled or running away. If this is done with a constant pressure drop $(p_0 - p_2)$ across the turbine the value of C_1 will be constant throughout the experiment and a graph of the same general shape as that shown in Figure 21.6 will be obtained by plotting the measured value of the work output W of the turbine against the rate of rotation ω of the rotor. With some additional measurements, the values of W and ω obtained from the coordinates of the maximum of this graph can be converted into values of η_{ideal} and U/C_1 respectively and the ratio of the two (which should be 4:1) determined.

Finally if a plot is made of ΔS against ω it will be found that maximum efficiency coincides with minimum increase in entropy, showing clearly the desirability of minimizing the irreversibility introduced in the turbine. This requirement is not immediately obvious from the application of the steady flow energy equation: the equation does not include the effects of irreversibility due, for example, to frictional heating. The heat term in the steady flow energy equation (which has been omitted earlier by considering the flow process to be adiabatic) refers only to heat flowing across the boundary of the system. Heat generated internally by friction is outside the scope of equation (21.1).

What happens can, however, be explained with the aid of an h—s diagram such as Figure 21.7. If the state of the air entering the turbine is represented by point A and the pressure drop across the turbine $(p_0 - p_2)$ is maintained constant, then the final state of the air leaving the turbine will be somewhere on the isobar corresponding to pressure p_2. If the flow process is adiabatic (i.e., no heat flow across the system boundary) and there is no internal friction the whole process is isentropic as represented by AC in Figure 21.7. This corresponds to the maximum possible work obtainable under adiabatic conditions when the pressure drop across the turbine is $(p_0 - p_2)$.

In actual practice some internal irreversibility is unavoidable and the maximum work output possible in practice will be obtained from a process represented by a line such as AB. When frictional heating occurs the consequent increase in entropy alters the final state of the air so that the value of the enthalpy at the outlet increases. If $(p_0 - p_2)$ is constant then the point representing outlet conditions moves along the p_2 isobar from C to B, resulting in a decrease in the overall enthalpy change, and hence of the work obtainable from the process. If B is taken to represent the state of the air leaving the turbine when its actual efficiency is a maximum, then as the internal irreversibility is increased by changing the speed of rotation the point B moves further along the p_2 isobar towards D. This reduces still more

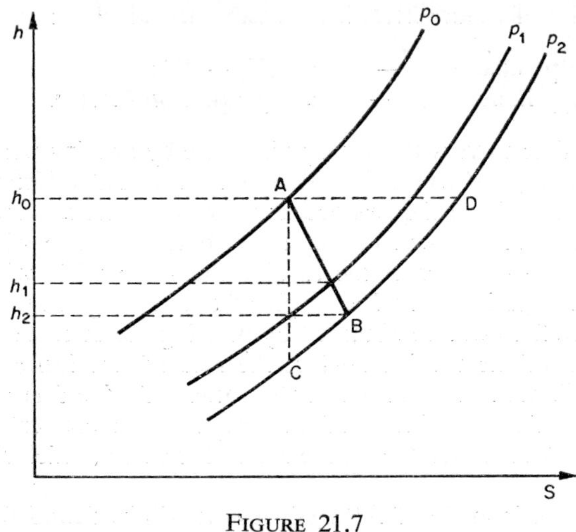

FIGURE 21.7

the value of the possible enthalpy change across the turbine and hence its work output. It follows at once that the maximum actual work output occurs in the case when the entropy change in the turbine is a minimum.

Experimental Details

Test Equipment. The inlet to the turbine (see Figure 21.8) is through a straight length of ducting with a cap C mounted over the central section of the fixed nozzle ring to give a uniform annular flow into the nozzles. The outlet is connected via an annular manifold directly to the suction fan. Rings of pressure tappings at p_0 and p_2 measure the inlet and outlet pressures respectively. An ordinary U-tube manometer will be adequate at p_2 which will give almost the full head developed by the fan. At p_0, however, the pressure will not be much less than atmospheric and an inclined manometer will have to be used for accurate measurements.

The mass flow through the turbine can conveniently be found in terms of the reading of the inlet pressure at p_0 making use of the effective annular venturi provided by the cap C. The turbine itself should be insulated as far as possible since the mass flow may not be high enough to make the assumption of adiabatic conditions valid.

Accurate measurement of the inlet and outlet temperatures (which are not very different) is difficult. One way of reducing the difficulty is to use two sets of thermocouples connected in series with their 'hot' junctions just projecting from small holes drilled in a half-tube inserted into the duct, with the leads, etc. shielded from the airflow (and still accessible) on the concave side of the tube. In this case the total voltage measured divided by the number of junctions gives a reasonably accurate value for the mean temperature of the flowing air. This method also makes it easy to measure the static temperature, though this is not strictly speaking necessary, as the flow velocity is

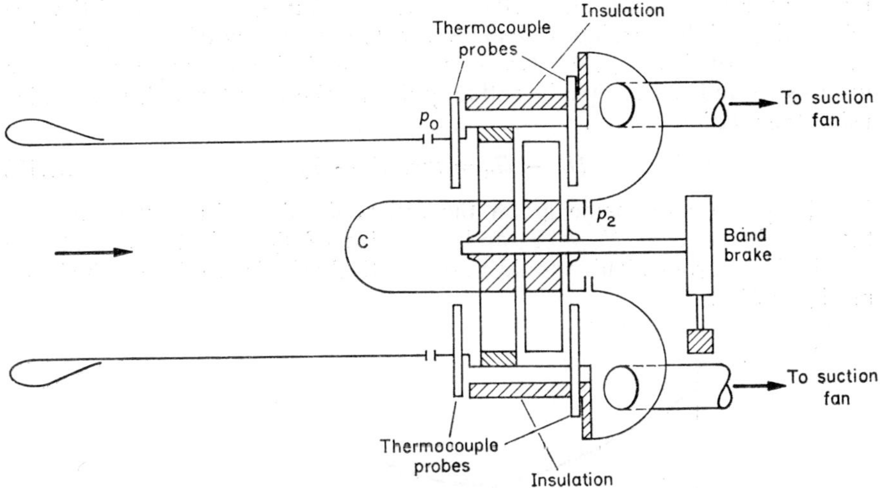

FIGURE 21.8

relatively low. It would be advisable to subject each junction to thermal cycling and to calibrate the junctions very carefully by inserting them, connected in series as they will be used, in a suitable constant temperature bath whose temperature is measured with a laboratory sub-standard thermometer. Another alternative is to use the same arrangement with the thermocouples connected individually, obtaining an average value of the temperature after examining the temperature profile. A third possibility is to measure the temperature difference between inlet and outlet with differential thermocouples and in addition to measure the absolute value of the inlet temperature.

The output of the turbine is easily measured by attaching to its axle a brake drum fitted with a band brake and to which suitable weights may be attached. The rate of rotation of the shaft has also to be measured; for this a simple tachometer should suffice.

Finally it is useful to have some control on the suction fan either by varying its speed or by using a throttling valve.

Procedure. When the apparatus is set up, first apply the brake sufficiently to prevent any rotation of the turbine shaft and adjust the fan to give a pressure drop across the turbine which can be maintained throughout the experiment. Measure the inlet and outlet pressure and temperature—the two temperatures should be equal. Next release the brake and maintaining the value of the pressure drop across the turbine, allow the turbine to 'run away'. Measure the rate of rotation of the turbine in this condition and also the values of inlet and outlet pressure and temperature. Again the two temperatures should be equal. Then apply varying brake loads to the turbine so that its rate of rotation increases systematically from zero to its value under run-away conditions. For each brake load measure the brake power produced by the turbine, the inlet and outlet pressures and temperatures, and the rate of rotation of the turbine. Calculate also the mass flow of the air through the turbine (which should have been constant during the experiment) and use this to find the change in enthalpy during the flow through the turbine using the relation

$$H_0 - H_2 = mc_p(T_0 - T_2) \tag{21.19}$$

Finally, plot on the same graph the observed value of the work output, the enthalpy change $H_0 - H_2$ and the value of T_2 against the rotational speed of the turbine. This should be qualitatively similar to the graph shown in Figure 21.9.

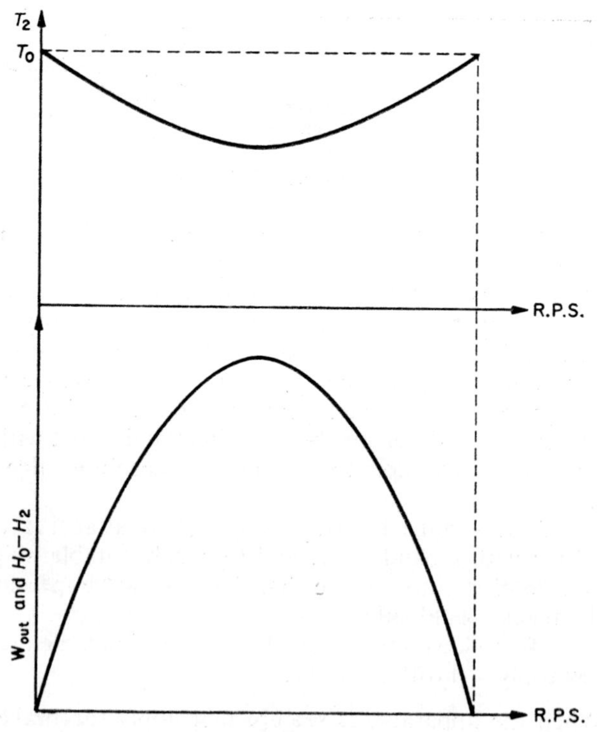

FIGURE 21.9

Further Experiment

The basic experiment can be extended by calculating C_1 and α_1 from the geometry of the nozzles and the measured value of the mass flow, if it is assumed that the air leaves the nozzles in a direction tangential to their surfaces. It is then possible to plot η_{ideal} *versus* U/c_1. In addition the actual entropy change can be calculated from equation (21.18) and also plotted against U/c_1. It is then possible to check that the ratio of the coordinates of the maximum point on the graph of η_{ideal} *versus* U/c_1 is 4:1.

References

1. Benson, R. S., Horlock, J. H. and Ryley, D, J., 'Introductory Experiments in Engineering Thermodynamics', *Bull. Mech. Eng. Educ.*, **1**, 9–19, 1962.
2. Rogers, G. F. C. and Mayhew, Y. R. *Engineering Thermodynamics: Work and Heat Transfer*, Longmans, 1970.

22

Direct Measurement of Absolute Thermodynamic Temperature

The absolute thermodynamic temperature T frequently appears in relations between thermodynamic variables such as the well known combined statement of the first and second laws of thermodynamics

$$T\,\mathrm{d}S = \mathrm{d}U + p\,\mathrm{d}V \tag{22.1}$$

From this it follows, as has already been shown (Experiment 14), that

$$\frac{1}{T} = \left(\frac{\partial S}{\partial U}\right)_V$$

If a relation of this type can be found in which the partial differentials involved represent measurable changes of thermodynamic properties, then absolute temperatures can be measured directly without needing a reversible heat engine on which calorimetric measurements are made.

Theory

Consider equation (22.1) rewritten in the form

$$\mathrm{d}U = T\,\mathrm{d}S - p\,\mathrm{d}V \tag{22.2}$$

Since $\mathrm{d}U$ is an exact differential it follows that

$$\left(\frac{\partial T}{\partial V}\right)_S = -\left(\frac{\partial p}{\partial S}\right)_V \tag{22.3}$$

Now, for a reversible process

$$\frac{\mathrm{d}Q}{T} = \mathrm{d}S \equiv \left(\frac{\partial S}{\partial p}\right)_V \mathrm{d}p + \left(\frac{\partial S}{\partial V}\right)_p \mathrm{d}V$$

since S can be written as a function of p and V and $\mathrm{d}S$ in an exact differential. Hence if V is constant,

$$\left(\frac{\partial S}{\partial p}\right)_V \mathrm{d}p = \left(\frac{\mathrm{d}Q}{T}\right)_{V=\text{const}}$$

$$\left(\frac{\partial S}{\partial p}\right)_V = \frac{1}{T}\left(\frac{\mathrm{d}Q}{\mathrm{d}p}\right)_{V=\text{const}}$$

$$\left(\frac{\partial p}{\partial S}\right)_V = T\left(\frac{\partial p}{\partial Q}\right)_V \tag{22.4}$$

Substituting in equation (22.3) into (22.4) we get

$$T = -\frac{(\partial T/\partial V)_S}{(\partial p/\partial Q)_V} \tag{22.5}$$

Thus T can be found directly by using a thermodynamic system which permits the two partial differentials in equation (22.5) to be evaluated separately when the system is in the same state of thermodynamic equilibrium (which of course is the state whose absolute thermodynamic temperature is T). This system temperature may be measured with an arbitrary practical thermometer whose scale of temperature gives a reading θ which in principle is to be arranged to be the same during the evaluation of both $(\partial T/\partial V)_S$ and $(\partial p/\partial Q)_V$.

Since to evaluate $(\partial T/\partial V)_S$ a temperature change has to be measured, this particular partial derivative can be measured only in the neighbourhood of θ where it will be safe to assume that the value of $(\partial T/\partial V)_S$ is effectively independent of temperature. Suppose the system starts initially at the temperature θ, chosen to be close to that of its surroundings, is compressed so that its temperature rises, and is then allowed to cool. If when the rate of cooling is stable the system is expanded rapidly by an amount ΔV (which is a little less than that of the initial compression), the temperature of the system will fall until it is approximately equal to the value θ. After thermal equilibrium has been re-established inside the system the temperature *versus* time curve will be an approximately horizontal line. This is shown in outline in Figure 22.1 where the line AB denotes the cooling until the change ΔV in volume occurs at B, and E′CD represents the subsequent temperature change. It is easy to extrapolate the line DC to E so that the length BE represents the change $\Delta\theta$ in the temperature of the system following an approximately isentropic change ΔV in its volume. Then if the relation between thermodynamic T and thermometer temperature θ is known,

$$\left(\frac{\partial T}{\partial V}\right)_S = \left(\frac{\Delta\theta}{\Delta V}\right)_S \left(\frac{\mathrm{d}T}{\mathrm{d}\theta}\right) \tag{22.6}$$

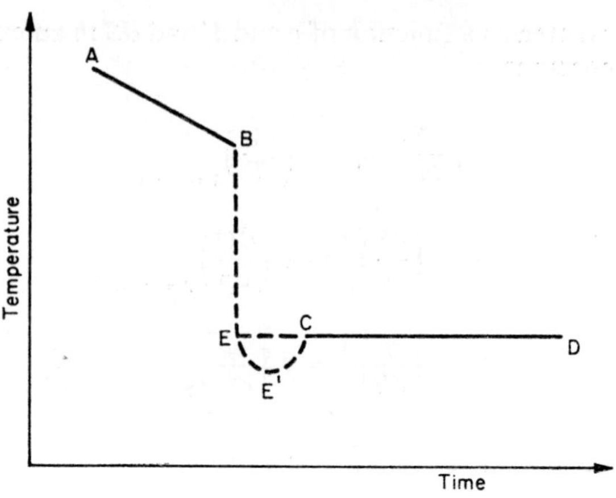

FIGURE 22.1

It is more straightforward to evaluate $(\partial p/\partial Q)_V$ since it is necessary only to measure the rate of increase in pressure at constant volume for a given heat input. Then,

$$\left(\frac{\partial p}{\partial Q}\right)_V = \left(\frac{\partial p}{\partial t}\right)_V \left(\frac{dQ}{dt}\right)_V^{-1} \tag{22.7}$$

where t denotes time. Now since some heat leakage is unavoidable, the value of $(dQ/dt)_V$ will represent the heat actually supplied to the system itself only at the instant when the system is at the temperature of its surroundings. Further the value of $(\partial p/\partial t)_V$ actually needed is that given by the tangent to the graph of p *versus* t at time $t = t_1$ (Figure 22.2) when the temperature of the system

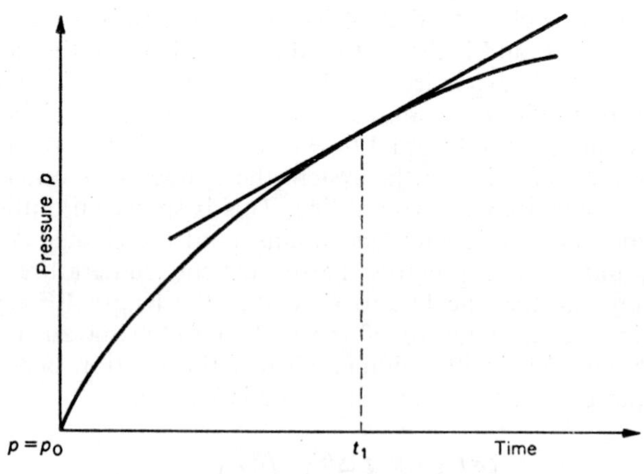

FIGURE 22.2

as measured by the thermometer chosen is θ. Using these two values equation (22.7) gives the appropriate value of $(\partial p/\partial Q)_V$. Then T can be found directly from equation (22.5).

It will be noted that in equation (22.6) a relation between T and θ is required to relate the temperature increment $\Delta\theta$ on the θ scale with the corresponding temperature increment ΔT on the absolute thermodynamic scale. This does not in principle presuppose the existence of a calibrated thermometer, though in practice one is necessary for the experiment described. From equation (22.5),

$$T = -\left(\frac{\partial T}{\partial V}\right)_S \left(\frac{\partial Q}{\partial p}\right)_V$$

$$= -\left(\frac{\Delta\theta}{\Delta V}\right)_S \left(\frac{\mathrm{d}T}{\mathrm{d}\theta}\right) \left(\frac{\partial Q}{\partial p}\right)_V$$

$$\frac{\mathrm{d}T}{T} = -\left(\frac{\Delta V}{\Delta\theta}\right)_S \left(\frac{\partial p}{\partial Q}\right)_V \mathrm{d}\theta$$

$$\ln\frac{T_2}{T_1} = -\int_{\theta_1}^{\theta_2} \left(\frac{\Delta V}{\Delta\theta}\right)_S \left(\frac{\partial p}{\partial Q}\right)_V \mathrm{d}\theta \tag{22.8}$$

The type of experiment described makes it possible in principle to evaluate thermometric as a function of θ the integrand of equation (22.8) which in turn leads to the ratio (T_2/T_1). If T_2 and T_1 are known such fixed points the difference $(T_2 - T_1)$ is a matter of definition. By making measurements at two such fixed points equation (22.8) gives the absolute thermodynamic temperature directly without relating $\Delta\theta$ and ΔT. This, however, is not feasible in practice with the apparatus described, which has a limited temperature range compared with the normal interval between thermometric fixed points.

Experimental Details

Test Equipment. The whole of the above theory depends on the actual process being effectively reversible so the apparatus must introduce as little irreversibility as possible. In addition very careful design is necessary to ensure that the measurements are viable.

The thermodynamic system chosen for the experiment is in essence a sealed metallic bellows containing a volatile liquid, a heating coil and a thermometer. In this system there is virtually no radial expansion due to internal pressure and external '$p\,\mathrm{d}V$' work may be written as $F\,\mathrm{d}L$ where F is the force applied to the bellows and $\mathrm{d}L$ is its change in length. Equation (22.2) then becomes

$$\mathrm{d}U = T\,\mathrm{d}S - F\,\mathrm{d}L$$

and analysis similar to that above gives

$$T = -\left(\frac{\partial T}{\partial L}\right)_S \left(\frac{\partial Q}{\partial F}\right)_L \tag{22.9}$$

External work is, of course, done against the atmosphere when the bellows expands but since measurements are required of increments in F only, the effect of the constant additional atmospheric load cancels. Changes in volume can be measured accurately using a clock gauge to measure the variation in the length of the bellows when a static load F is applied to it.

A suitable bellows [1] is a brass one with 12 convolutions, 47 mm in diameter, of unstrained length 37 mm, and capable of withstanding a maximum internal pressure of 17 atmospheres. Such a bellows has a maximum range of extension of 10 mm. There is a danger of introducing hysteresis and, hence, irreversibility if extensions approaching the maximum value are used; for this reason stops are fitted to the apparatus to limit the extension to about 5 mm. With such a bellows partially filled with either sulphur dioxide or dichlorodifluoromethane operating at about 40 °C the temperature change observed for a 5 mm change in the length of the bellows will be about 0·25 °C.

To provide the insulation needed the bellows X is supported between two glass tubes G_1 and G_2 as shown in Figure 22.3. The top of the bellows is closed by a copper cap C, machined to locate the upper glass insulator, and carrying a closed copper cylinder through the base of which passes a thermometer pocket N, soldered in position. Inside this cylinder is a heating coil H. Leads to this coil and to the thermometer M are connected, via ceramic-to-metal seals mounted in the flange O and the cap C, to suitable voltage supplies. The thermometer used may be either a resistance thermometer or a carefully made thermocouple. In either case the leads to it must be thin to reduce heat losses by conduction along them.

The lower end of the bellows is sealed by another cap D with a central valve I permitting the introduction into the bellows of the chosen fluid.

The upper glass insulator G_1 is located in a brass flange J connected by a tube Y to a second flange K which clears the lower insulator G_2. This is located on a flange R provided with three holes through which rods connect to K a disc A. At the centre of A is a rod E on which weights may be hung. This rod passes through a hole in the base B of the apparatus, the hole being sealed with an O-ring. Attached to B is a large diameter tube Z closed by a flange O which completely encloses the apparatus. Supports between O and B carry wheels W to act as guides which ensure that the expansion of the bellows is axial. A connection from O enables the whole apparatus to be evacuated. A heating coil on the outside of tube Z enables the temperature of the surroundings of the bellows to be maintained at some suitable constant value. It is not essential to evacuate the surroundings of the bellows system but it is desirable to do so since this approximately halves the heat losses which are of the order 0·06 watts per degree without the vacuum.

Procedure. First allow the system to reach thermal equilibrium with its surroundings with the bellows X at its maximum length. Then quickly load the bellows to give its maximum permitted compression by adding weights to E (the maximum compression will require a mass of the order 100 kg). This

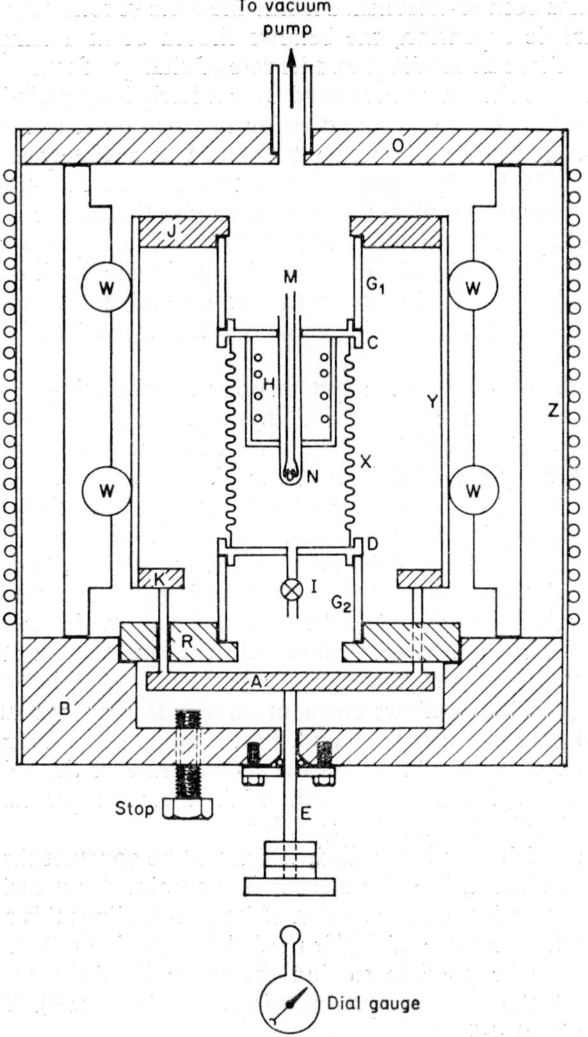

FIGURE 22.3

compression will raise the temperature inside the bellows and so, to compensate for heat losses before the measurement begins, supply a little additional heat using the coil H. Plot a cooling curve, adjusting the load on E during the cooling so that the length of the bellows remains constant at some arbitrarily chosen value; this ensures that no extra work is done on the bellows system by the load as the bellows cools and hence that the temperature change with time is entirely due to cooling. When the cooling curve is linear rapidly reduce the force on the bellows by removing sufficient weights from E for the bellows to expand to its original length. After thermodynamic equilibrium has been re-established inside the bellows following this approximately isentropic expansion, the bellows should be at a temperature very close to that of its surroundings and its rate of change of temperature should be very low. This should be measured over a fairly long period (of approximately $\frac{1}{2}$ to 1 hour), meanwhile adjusting the load on E to keep the length of the bellows constant at the value it assumed immediately after its sudden increase in length. Finally a graph similar to Figure 22.1 should be obtained and it will be easy to extrapole the line DC to give the change $\Delta\theta$ in temperature measured by the thermometer M following an isentropic expansion whose magnitude should be found from the change in the readings of the dial gauge. The value of $(\partial T/\partial L)_S$ at the temperature of the surroundings of the bellows is then readily found using equation (22.6), with L taking the place of V.

The next stage is to cool the system prior to measuring $(\partial p/\partial t)_V$. This is achieved by breaking the vacuum (if any) and cooling the bellows directly using a jet of cold air which can be directed on to the bellows through a small hole specially provided in the tube Y. When the temperature of the bellows as measured by M has fallen by about 10 °C, reseal the whole system and evacuate it as quickly as possible. Then leave it for at least thirty minutes to allow the surroundings of the bellows, i.e. the tube Y, to attain thermal equilibrium with the enclosing tube Z, and during this period adjust the load on E to maintain the length of the bellows constant and in the middle of the range available. Next heat the bellows via the coil H at a steady rate of about 0·5 watt. During this heating increase the load on E in steps of 1 kg to keep the length of the bellows constant, and note the times at which each weight is added to the nearest second. Plot a graph of load *versus* time and measure its slope at the time when the temperature of the bellows system, as measured by M, is the same as that of the mean temperature during the earlier approximately isentropic expansion.

The rate at which heat is supplied, which will be constant throughout this part of the experiment, is easily measured. As pointed out earlier when the system temperature is the same as that of its surroundings this rate of supply of heat is exactly the value of $\mathrm{d}Q/\mathrm{d}t$ needed in equation (22.7).

It should now be possible to find the value T of the thermodynamic temperature of the bellows system using equation (22.9). The accuracy attainable is about ±5 °C.

Reference

1. Bancroft, D. 'Calorimetric Determination of Absolute Temperature', *Amer. J. Phys.* **23,** 142, 1955.

23

Measurement of Stefan's Constant

Although thermodynamics originated in the study of ways of improving our understanding of the operation of heat engines, the fundamental nature of the theory subsequently developed is illustrated by our ability to apply thermodynamic arguments to many other situations. One such case is the study of radiation, where concise thermodynamic arguments readily lead to the form of the relation between the quantity of energy radiated by a body and its absolute temperature, i.e. Stefan's law. The value of Stefan's constant cannot be derived analytically without appealing to quantum statistical thermodynamics. However, it can be measured fairly easily provided it is possible to make a surface which approximates to an ideal black surface with an absorptivity of unity. Alternatively the emissivities of particular types of surface may be measured if a value of Stefan's constant is assumed.

Theory

The radiation inside any enclosure which is in thermal equilibrium at an absolute temperature T is known to be black body radiation corresponding to the temperature T. The photons which make up this radiation move about at random inside the container, making collisions which may be assumed to be effectively elastic with the walls. Under these conditions it can be shown (see Appendix 23A) that the radiation pressure p inside the enclosure is given by the relation

$$p = \tfrac{1}{3}u \tag{23.1}$$

where u is the energy density of the photons inside the enclosure, whose internal volume may be taken to be V. Then, if U denotes the internal energy of the thermodynamic system represented by the interior of the enclosure,

$$U = Vu \tag{23.2}$$

Using the first and second laws of thermodynamics applied to a quasistatic expansion of the system inside V, assuming that there is no mutual interaction between photons, gives

$$\begin{aligned} T\,\mathrm{d}S &= \mathrm{d}U + p\,\mathrm{d}V \\ &= V\,\mathrm{d}u + u\,\mathrm{d}V + \tfrac{1}{3}u\,\mathrm{d}V \quad \text{(from 23.1 and 23.2)} \\ &= V\,\mathrm{d}u + \tfrac{4}{3}u\,\mathrm{d}V \end{aligned}$$

Hence $$\mathrm{d}S = \frac{V}{T}\,\mathrm{d}u + \frac{4}{3T}u\,\mathrm{d}V \qquad (23.3)$$

Since $\mathrm{d}S$ is a perfect differential,

$$\frac{\partial}{\partial V}\left(\frac{V}{T}\right)_u = \frac{\partial}{\partial u}\left(\frac{4u}{3T}\right)_u$$

i.e. $$\frac{1}{T} = \frac{4}{3T} - \frac{4u}{3T^2}\frac{\partial T}{\partial u}$$

which may be reduced to

$$\frac{\mathrm{d}u}{u} = 4\frac{\mathrm{d}T}{T}$$

Thus $$u = \text{const.} \times T^4 \qquad (23.4)$$

Radiation theory shows that if E is the total radiant energy emitted in one second into a hemisphere on one side of a black surface of unit area which forms part of an enclosure in which the energy density is u, then

$$u = 4E/c \qquad (23.5)$$

where c is the velocity of light. Hence, combining equations (23.4) and (23.5),

$$E = \sigma T^4 \qquad (23.6)$$

where σ is Stefan's constant.

Consider the radiation inside the hemispherical enclosure illustrated in Figure 23.1, formed by the non-conducting plate A and the metal hemisphere B. If A and B are both at temperature T so that the enclosure is in thermal equilibrium, the quantity of radiation falling in unit time on any small area of the enclosing walls is equal to the quantity of radiation it emits in the same period. Suppose a hole is drilled in the plate A and a small metal disc D inserted as shown. When the enclosure is in equilibrium at temperature T the radiant energy absorbed by D equals the radiant energy it emits, and this is given by

$$E = \sigma A_D T^4 \qquad (23.7)$$

where A_D is the area of the disc D, which is assumed to have an emissivity of unity, and σ is Stefan's constant.

Now suppose the disc D is removed from the plate A. If the area of the disc is a small fraction of the total area of the walls of the hemispherical enclosure the thermal equilibrium of the interior of this enclosure will not be altered significantly and the radiant energy falling on the area previously occupied by D will be unchanged and given by equation (23.7).

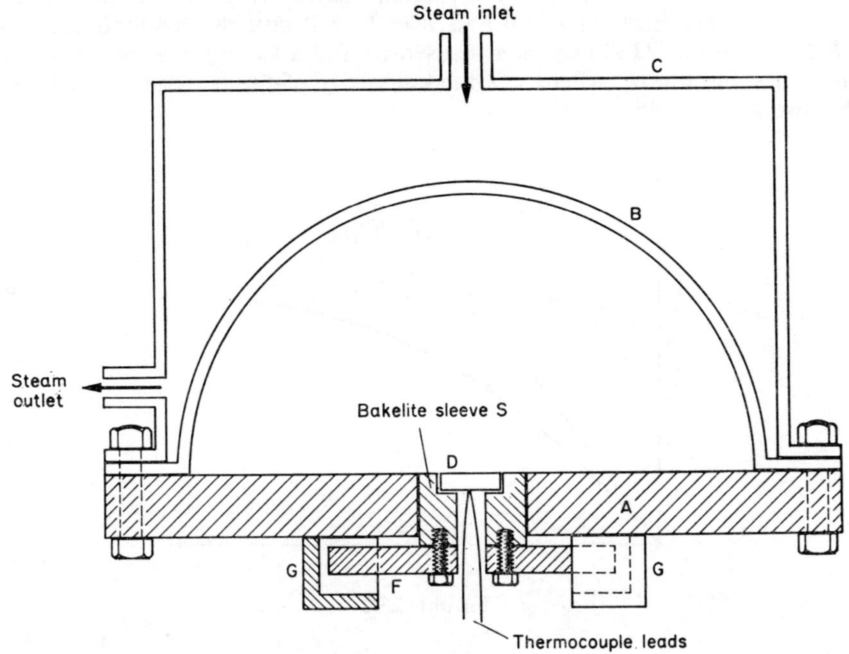

FIGURE 23.1

If now the disc D is re-inserted in A at a time when the temperature of D is T_1 then the radiant energy it emits into the enclosure in unit time will be given by

$$E_1 = \sigma A_D T_1^4 \tag{23.8}$$

The radiation absorbed will still be given by equation (23.7) so if $T > T_1$ there will be a net heat input to D in unit time given by

$$E - E_1 = \sigma A_D(T^4 - T_1^4)$$

If the disc D has mass m and specific heat s (measured in energy units) then a short time after D is inserted in A,

$$ms\frac{dT}{dt} = \sigma A_D(T^4 - T_1^4)$$

$$\sigma = \frac{ms\, dT/dt}{A_D(T^4 - T_1^4)} \tag{23.9}$$

In this equation dT/dt denotes the rate of rise of temperature of the disc D at the instant when its temperature is T_1 and will vary with the value of T_1. It is clearly best measured at time $t = 0$ before heat conducted from A to D begins to have any significant effect and when there is very little uncertainty about the actual value of T_1. If T is plotted against t (Figure 23.2) the tangent to the resulting curve at $t = 0$ and $T = T_1$ will give the required value of dT/dt. Equation (23.9) can then be used to find a value for σ, or more precisely a value for $\varepsilon\sigma$, where ε is the emissivity of the material from which the disc D was made.

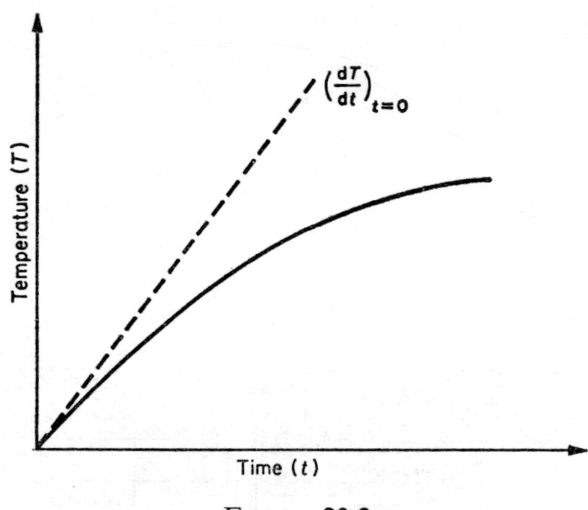

FIGURE 23.2

Experimental Details

Test Equipment. The apparatus, illustrated in Figure 23.1, is centred on a flanged copper hemisphere B fixed to a flat non-conducting plate A. The outer surface of B is enclosed in a metal steam or water jacket used to heat B to some suitable constant temperature. The hemispherical shape for B is chosen solely on the grounds that it simplifies the task of draining condensed steam from the space between B and C. Further, by directing incoming steam to strike B symmetrically at its highest point, the distribution of steam is reasonably uniform. A mercury-in-glass thermometer is used to measure the temperature inside the enclosure and several thermocouples are attached to various points on the surface of B to measure its mean temperature. These thermocouples must, of course, be capable of immersion in water; alternatively they could be attached to the inner surface of B and the wires brought out through the base A.

The disc D, which should be about 2 cm in diameter and 0·25 cm thick, is mounted in an insulating bakelite sleeve S which fits the hole drilled in the middle of A. The base of S is bolted to a flat bar F, of any convenient material (metal if necessary) which is used to lock S and D in position by locating F in two supports G on the underside of A.

The leads of a thermocouple attached to D are brought out through an axial hole in S and taken to a cold junction. This is constructed by joining both thermocouple leads to copper wires and encapsulating both these junctions in a cylinder of paraffin wax, which is inserted in a freezing mixture in a Dewar flask. The paraffin wax insulates the junctions and the leads from the freezing mixture and so permits the use of thermocouple leads with sleeving not impervious to water.

The two copper leads coming from the cold junction are attached to a sensitive moving-coil voltmeter or some other sensitive fast acting instrument such as a digital voltmeter, or a constant-resistance deflectional potentiometer. It should be borne in mind that the voltage to be measured is changing with time so that, unless the response time of the measuring device is very short, the dynamics of the chosen instrument may affect the values obtained. (Appendix 23A.2 considers the response of a moving coil galvanometer.)

Procedure. It will be assumed that the apparatus is steam heated. This is by no means essential as the thermal inertia of hot water filling the space between B and C is more than adequate to prevent significant cooling in the time required for taking readings; furthermore it ensures more uniform heating of the hemisphere B than is possible with steam.

Before the apparatus is assembled the inside surfaces of the hemispherical enclosure bounded by B and A should be blackened to make their absorptivity approximately unity. The upper surface of D should also be blackened. This blackening is very conveniently accomplished using burning camphor to produce a suitable deposit of carbon on these surfaces.

Then with the disc D removed from the hole in A, heat the whole apparatus by connecting a steam supply to the connection on the top of the steam jacket. Leave the apparatus for some time until its temperature is uniform so that inside the hemispherical space the radiation approximates closely to black body radiation. This process occurs fairly quickly if the whole of the apparatus above the plate A is insulated to reduce heat losses.

Quickly place the disc D, with its insulating sleeve S, in position by locking the bar F in the slots on the supports G, and at the same time start a stop-watch. Take readings of the temperature at short time intervals. This process need not continue for more than 20 seconds since all that is needed is to construct enough of the graph of T *versus* t to permit the measurement of the slope of this graph at $t = 0$. Furthermore, the longer D is left in position the greater is the probability of errors due to heat conduction from A to D. After a sufficient number of readings have been taken, remove the disc D and allow it to cool; then repeat the whole process.

Finally calibrate the thermocouple attached to D. This is best done by immersing D in oil in a suitable glass vessel which is then heated in a water bath until the thermal e.m.f. produced is slightly greater than the maximum value measured during the experiment. Take care to stir the oil throughout.

When the calibration has been completed, convert the thermocouple readings obtained in the experiment into temperatures and plot a graph showing the rate at which the temperature of D increases. Then use equation

(23.9) to calculate σ, Stefan's constant, on the assumption that the emissivity of the surface of D is unity. Alternatively, the emissivity of the surface of D may be calculated using the known value of σ.

Appendix 23A.1—Radiation Pressure

The existence of pressure due to radiation is a necessary consequence of the fact that radiation in an enclosure exhibits the property of energy density. If no radiation pressure was exerted on the walls it would be possible to compress the radiation and increase its energy density and hence its temperature without doing work.

To derive an expression for the magnitude of the pressure due to radiation the simplest method is to regard isotropic radiation as composed of a large number of non-interacting photons. Each photon has energy $h\nu$ (where h is Planck's constant and ν is the frequency of the radiation) and momentum $h\nu/c$ (where c is the velocity of light). Suppose there are n photons present in unit volume and that the distribution function for the number of photons with frequency ν is the black-body distribution.

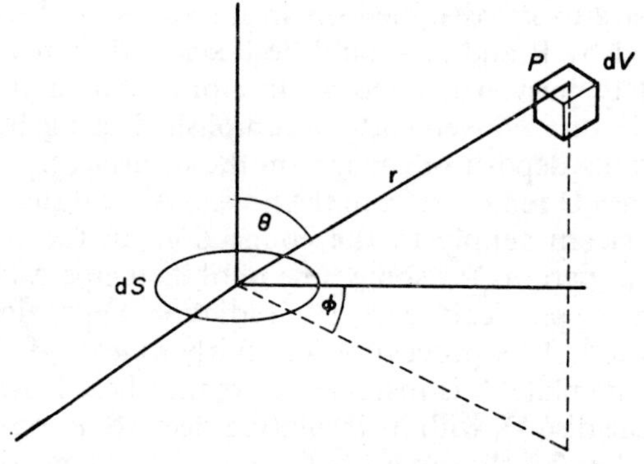

FIGURE 23.3

Consider a volume $\mathrm{d}V$ centred on the point P whose coordinates are (r, θ, ϕ). Let the number of photons in unit volume with energies between $h\nu$ and $h(\nu + \mathrm{d}\nu)$ be $n_\nu\, \mathrm{d}\nu$. Then $\mathrm{d}V$ contains $n_\nu\, \mathrm{d}\nu\, \mathrm{d}V$ such photons where $\int_0^1 \mathrm{d}V \int_0^\infty n_\nu\, \mathrm{d}\nu = n$, the total number of photons in unit volume.

Of the radiation leaving $\mathrm{d}V$ only a fraction $\mathrm{d}\Omega/4\pi$ will reach the small area $\mathrm{d}S$ centred on the origin where $\mathrm{d}\Omega$, the solid angle subtended at P by $\mathrm{d}S$, is given by

$$\mathrm{d}\Omega = \frac{\mathrm{d}S \cos\theta}{r^2}$$

Therefore the number of photons with energy in the range $h\nu$ to $h(\nu + d\nu)$ which originate in dV and strike dS in time dt is

$$n_\nu \, d\nu \, dV \frac{dS \cos \theta}{4\pi r^2} \tag{23A.1.1}$$

where $dV = r^2 \sin \theta \, dr \, d\theta \, d\phi$ in spherical polar coordinates.

The total number of photons striking dS in time dt will be given by integrating equation (23A.1.1) over a hemisphere of radius $c\,dt$ (where c is the velocity of light) since all the photons which succeed in striking dS in dt must originate inside this hemisphere. Choosing dt so that the product $c\,dt$ is less than the dimensions of the enclosure, the total number of photons striking the area dS in time dt

$$\int_{\nu=0}^{\infty} \int_{r=0}^{c dt} \int_{\theta=0}^{\pi/2} \int_{\varphi=0}^{2\pi} n_\nu \, d\nu \, \frac{dS \cos \theta}{4\pi r^2} \, r^2 \sin \theta \, dr \, d\theta \, d\phi$$

At each impact for which the angle of incidence is θ the momentum change, assuming perfectly elastic collisions, is

$$\frac{2h\nu}{c} \cos \theta$$

Hence the radiation pressure p is given by

$$p = \frac{1}{dSdt} \int_{\nu=0}^{\infty} \int_{r=0}^{c dt} \int_{\theta=0}^{\pi/2} \int_{\varphi=0}^{2\pi} n_\nu \, d\nu \frac{2h\nu}{c} \cos \theta \, . \, \frac{dS \cos \theta}{4\pi r^2} \, . \, r^2 \sin \theta \, dr \, d\theta \, d\phi$$

$$= \frac{1}{dSdt} \frac{2}{c} \frac{dS}{4\pi} \int_0^{\infty} n_\nu (h\nu) \, d\nu \int_0^{c dt} dr \int_0^{\pi/2} \cos^2 \theta \sin \theta \, d\theta \int_0^{2\pi} d\phi$$

$$= \tfrac{1}{3} \int_0^{\infty} n_\nu (h\nu) \, d\nu$$

Now this integral represents the total energy in unit volume i.e. the energy density u. Therefore

$$p = \tfrac{1}{3} u$$

This derivation is, of course, analogous to that used to derive an expression for the pressure of an ideal gas.

Appendix 23A.2—Galvanometer dynamics

Suppose a steady temperature difference of 1 °C between the hot and cold junctions of a thermocouple gives a thermal e.m.f. which will cause a galvanometer deflection of β radians. Then the deflection θ_1 due to a steady temperature difference of T_1 °C is given by

$$\theta_1 = \beta T_1 \tag{23A.2.1}$$

If μ denotes the torsional couple applied by the galvanometer suspension per radian deflection, the equation of motion of the galvanometer coil at time t

when its deflection is θ and the temperature difference between the hot and cold junctions of the thermocouple is T is

$$\mu\beta T - \mu\theta = I\frac{d^2\theta}{dt^2} + K\frac{d\theta}{dt} \qquad (23A.2.2)$$

where I is the moment of inertia of the coil and

K is the damping coefficient for the suspended system.

Writing $K/I = k$ and $\mu/I = n^2$, this equation becomes

$$\frac{d^2\theta}{dt^2} + k\frac{d\theta}{dt} + n^2\theta = n^2\beta T \qquad (23A.2.3)$$

If T approaches a steady value T_0 in accordance with the equation

$$T = T_0(1 - e^{-pt}) \qquad (23A.2.4)$$

where p is a constant, equation (23A.2.3) becomes

$$\frac{d^2\theta}{dt^2} + k\frac{d\theta}{dt} + n^2\theta = n^2\beta T_0(1 - e^{-pt}) \qquad (23A.2.5)$$

The solution of this equation is

$$\theta = \beta T_0\left(1 - \frac{n^2e^{-pt}}{p^2 - kp + n^2}\right) - Ae^{-kt/2}\sin\left[\left(n^2 - \frac{k^2}{4}\right)^{\frac{1}{2}}t + \alpha\right] \qquad (23A.2.6)$$

where A is a constant and α is a phase angle.

The second term is oscillatory if $n^2 > \frac{1}{4}k^2$ but may be neglected if k is large enough for this term to decay rapidly. The electromagnetic damping of most galvanometer coils is fairly large and the value of I is usually small so it is reasonable to assume that k is large enough for the second term in equation (23A.2.6) to be neglected after a short time has elapsed. Making this assumption equation (23A.2.6) becomes

$$\theta = \beta T_0\left(1 - \frac{n^2e^{-pt}}{p^2 - kp + n^2}\right) \qquad (23A.2.7)$$

and hence

$$\frac{d\theta}{dt} = \beta pT_0e^{-pt}\frac{n^2}{p^2 - kp + n^2} \qquad (23A.2.8)$$

From equation (23A.2.4)

$$\frac{dT}{dt} = pT_0e^{-pt} \qquad (23A.2.9)$$

Since only the value of the initial rate of rise of temperature is important,

$$\left(\frac{dT}{dt}\right)_{t=0} = pT_0 \qquad (23A.2.10)$$

and from equation (23A.2.8)

$$\left(\frac{\mathrm{d}\theta}{\mathrm{d}t}\right)_{t=0} = \frac{\beta p T_0 n^2}{p^2 - kp + n^2} \tag{23A.2.11}$$

$$\left(\frac{dT}{\mathrm{d}t}\right)_{t=0} = \frac{p^2 - kp + n^2}{n^2}\left(\frac{1}{\beta}\right)\left(\frac{\mathrm{d}\theta}{\mathrm{d}t}\right)_{t=0} \tag{23A.2.12}$$

It is easy to find the value of β by calibrating the thermocouple directly in terms of the galvanometer deflection. A graph of θ *versus* T over the small range of interest will certainly be linear and its slope is β.

By plotting θ *versus* t for the hot junction, $(\mathrm{d}\theta/\mathrm{d}t)_{t=0}$ may be found by measuring the slope of this graph at $t = 0$. Then the only unknown in equation (23A.2.12) whose value is needed to calculate $(\mathrm{d}T/\mathrm{d}t)_{t=0}$ is the term

$$\frac{p^2 - kpn^2}{n^2}$$

The value of n. When the galvanometer coil is open-circuited there is no electromagnetic damping and the value of k is negligible. In this case the period P of the oscillations will be given by equation (23A.2.6) with $k = 0$. This then gives

$$n = \frac{2\pi}{P} \tag{23A.2.13}$$

The value of k. When the galvanometer coil is swinging with the external circuit closed but with no input from the thermocouple, so that T_0 is zero, equation (23A.2.6) gives

$$\theta = -Ae^{-kt/2}\sin\left[\left(n^2 - \frac{k^2}{4}\right)^{\frac{1}{2}} t + \alpha\right]$$

This is the equation of damped simple harmonic motion. If θ_1, θ_2, θ_3 and θ_4 are the observed maximum values of the first four successive deflections and if P_0 is the period of the oscillation

$$\theta_1 = -Ae^{-k/2\ P_0/4}\sin\left(\frac{2\pi}{P_0}\frac{P_0}{4} + \alpha\right)$$

$$\theta_2 = -Ae^{-k/2\ 3P_0/4}\sin\left(\frac{2\pi}{P_0}\frac{3P_0}{4} + \alpha\right) \qquad \text{etc.}$$

Hence

$$\theta_1 - \theta_2 = Ae^{-k/2\ P_0/4}\sin\left(\tfrac{1}{2}\pi + \alpha\right)[e^{-P_0/2} - 1]$$

Similarly

$$\theta_3 - \theta_4 = Ae^{-k/2\ 5P_0/4}\sin\left(\tfrac{1}{2}\pi + \alpha\right)[e^{-P_0/2} - 1]$$

Hence

$$\log\frac{\theta_1 - \theta_2}{\theta_3 - \theta_4} = \frac{kP_0}{2} = \frac{k\pi}{(n^2 - \frac{1}{4}k^2)^{\frac{1}{2}}} \tag{23A.2.14}$$

The value of p_1. At time t_1 equation (23A.2.8) gives

$$d\theta_1/dt = Bpe^{-pt_1}$$

where

$$B = \frac{\beta n^2 T_0}{p^2 - kp + n^2} = \text{a constant}$$

Similarly at time $t_2 = 2t_1$

$$\frac{d\theta_2}{dt} = Bpe^{-2pt_1}$$

Hence

$$\left(\frac{d\theta_1}{dt}\right) \Big/ \left(\frac{d\theta_2}{dt}\right) = e^{pt_1}$$

and

$$\log\left[\left(\frac{d\theta_1}{dt}\right) \Big/ \left(\frac{d\theta_2}{dt}\right)\right] = pt_1 \qquad (23A.2.15)$$

Hence if a graph is plotted of galvanometer deflection θ *versus* time the slopes of the graph measured at times t_1 and $2t_1$ substituted in the last equation will enable the value of p to be found.

Part III

Heat Transfer

Introduction

Heat transfer occurs when energy is passed from one location, the source, to another, the receiver, as a result of a temperature difference between them. In the absence of heat leaks the heat transfer from the source equals that to the receiver. Also the heat transfer occurs from the higher temperature source to the lower temperature receiver but never spontaneously in the reverse direction.

The engineer is required to deal effectively and economically with many processes which involve heat transfer. Often the aim is to reduce the resistance to heat transfer, so reducing the size, weight and cost of the plant necessary to perform a particular heat-transfer duty. Sometimes the aim is to increase the resistance to heat transfer, i.e. to provide thermal insulation at a cost less than the cost of the heat energy saved thereby. In either case the concern is to know heat transfer rates. Occasionally this knowledge may be obtained by analytical solutions of algebraic or graphical type. The analytical approach usually requires the situation to be idealized, perhaps to an extent that a solution of sufficient accuracy cannot be achieved. Often an experimental investigation on the plant concerned, or perhaps on some analogous model of it, is necessary. The experimental approach essentially involves objective physical measurement. Normally temperature-measuring techniques are used but there are limits to the accuracy which can be achieved. Hence both analytical approaches require judgement which can be developed by experience.

Heat transfer occurs by three means: convection, conduction and radiation.

Convection. Heat transfer can occur within a gas, vapour or liquid due to fluid motion. This motion arises from density changes caused by temperature gradients within the fluid and is termed natural or free convection. In many situations an overall economy is achieved by increasing the heat transfer rate as a result of increasing the fluid motion by mechanical means such as a fan or pump; this is termed forced convection. The other kinds of heat transfer are invariably present during heat transfer by convection but in many engineering applications are relatively negligible, particularly under forced convection conditions. Convection heat transfer between a fluid and a solid or liquid surface may be accompanied by a transfer of mass. Common examples are the condensation of vapours and evaporation of liquids.

Conduction. Higher-temperature particles of matter have higher energy levels than adjacent lower-temperature particles, and heat energy can be transferred in gases and liquids by a complex mechanism of molecular collision or in solids by lattice-vibration propagation. Also, in metals the flow of free electrons greatly enhances heat transference by conduction. Conduction in solids and liquids involves intermolecular energy transfer without the molecular translation associated with convection.

Radiation. This is the transfer of energy due to an electromagnetic wave phenomenon. Unlike heat transfers by convection and conduction, which depend on temperature differences between the source and receiver, the intensity of radiation emitted depends only on the temperature of the source. Since no source or receiver is at zero absolute temperature there is a continual radiation exchange between them. Radiation heat transfer does not necessarily increase the temperature of the intervening medium. A medium is not even necessary; hence energy from the sun is transmitted through outer space to the earth. A fraction of the radiation incident on a receiver is absorbed, a fraction is reflected and a fraction may be transmitted through the receiver. The magnitude of each fraction depends greatly on whether the receiver is a gas, a liquid or a solid, but the sum of the three fractions must equal unity.

With few exceptions, engineering problems involve more than one of the three kinds of heat transfer. Their solution requires the application of laws governing each kind. Although the ultimate interest may be in the total heat transfer from the source to the receiver, each kind must be considered separately because, even with fixed temperatures at the source and at the receiver, an important factor may be changed which affects only one of the kinds involved. The following experiments are designed to confirm and apply the relevant laws and to convey the essential nature of each heat transfer process.

24

Heat Transfer by Convection

The heat transfer process in gases and liquids is affected by the mechanics of fluid flow in addition to thermodynamic phenomena, so there are many variables involved. Much time would be required to determine experimentally the relationship between individual variables, but by dimensional analysis they can be arranged in a smaller number of dimensionless groups. If correlations between the smaller number are found much time is saved and presentation of the experimental results is thereby greatly simplified. The arrangement of the relevant variables into groups can be achieved in simple situations by an understanding of the physical processes involved or, in a complex case, such as that of convective heat transfer, by the algebraic approach known as the method of indices. Initially it is necessary to know, or to make an intelligent guess at, the variables likely to affect the process.

Theory

Those variables relevant to the heat transfer at a surface by convection are listed in Table 24.1 together with their symbols and fundamental dimensions.

The surface heat transfer coefficient h is the factor of proportionality in the relationship known as the Newton equation which states that the heat transfer rate per unit area of surface $\dot{q}$ is proportional to the temperature difference θ between the surface and a fluid in contact with it.

It is assumed that the dependent variable $\dot{q}$ (or h) is a function of the independent variables l, θ, v, μ, ρ, c and k and of the product βg, where βg is proportional to the change of fluid buoyancy against gravitational force. By substituting the basic dimensions chosen in Table 24.1 this statement may be written as

$$\left(\frac{\mathrm{Q}}{\mathrm{TL}^2}\right) = \text{constant }(\mathrm{L})^A(\theta)^B\left(\frac{\mathrm{L}}{\mathrm{T}}\right)^C\left(\frac{\mathrm{M}}{\mathrm{LT}}\right)^D\left(\frac{\mathrm{M}}{\mathrm{L}^3}\right)^E\left(\frac{\mathrm{Q}}{\mathrm{M}\theta}\right)^F\left(\frac{\mathrm{Q}}{\mathrm{LT}\theta}\right)^G\left(\frac{\mathrm{L}}{\theta\mathrm{T}^2}\right)^H \quad (24.1)$$

Since this is an equation for a physical system it must be dimensionally homogeneous. Hence the indices may be equated, giving five equations containing eight unknowns. Solving in terms of, say, C, E, and F and grouping,

$$\left(\frac{\dot{q}l}{k\theta}\right) = \text{constant }\left(\frac{\rho vl}{\mu}\right)^E\left(\frac{l\beta g\theta}{v^2}\right)^{(E-C)/2}\left(\frac{c\mu}{k}\right)^F \quad (24.2)$$

TABLE 24.1 *Symbols and Dimensions of Variables*

Variable	*Symbol*	*Exponents of units of*				
		Heat Q	*Mass* M	*Length* L	*Time* T	*Temperature* θ
Heat transfer rate at surface per unit area	$\dot{q}$	1	0	−2	−1	0
Surface heat transfer coefficient	h	1	0	−2	−1	−1
Size of configuration	l	0	0	1	0	0
Temperature difference across surface	θ	0	0	0	0	1
Velocity of fluid	v	0	0	1	−1	0
Fluid viscosity at bulk temperature	μ	0	1	−1	−1	0
Fluid density	ρ	0	1	−3	0	0
Fluid specific heat	c	1	−1	0	0	−1
Fluid thermal conductivity	k	1	0	−1	−1	−1
Fluid coefficient of cubical expansion	β	0	0	0	0	−1
Acceleration due to gravity	g	0	0	1	−2	0

Buoyancy effects activate the fluid motion when heat transfer is by natural convection. The group including βg must then be independent of v and this can be achieved by multiplying it by another (dimensionless) term $(\rho vl/\mu)^2$. Using single symbols, x, y and z for the indices, and since $\dot{q}/h = \theta$ from the Newton equation then

$$\left(\frac{hl}{k}\right) = \text{constant}\left(\frac{\rho vl}{\mu}\right)^x \left(\frac{l^3\beta g\theta\rho^2}{\mu^2}\right)^y \left(\frac{c\mu}{k}\right)^z$$

or

$$\text{Nu} = \text{constant Re}^x\,\text{Gr}^y\,\text{Pr}^z \tag{24.3}$$

where Nu, Re, Gr, and Pr are symbols for the corresponding dimensionless groups and are known as the Nusselt, Reynolds, Grashof and Prandtl numbers respectively.

For natural convection,

$$\text{Nu} = \text{constant Gr}^y\,\text{Pr}^z \tag{24.4}$$

For forced convection,

$$\text{Nu} = \text{constant Re}^m\,\text{Pr}^n \tag{24.5}$$

The constants and indices in equations such as (24.4) and (24.5) can be determined only by experiment. In equation (24.4), since $v \doteqdot 0$, the Reynolds number is omitted. In equation (24.5), where $v \neq 0$, the Reynolds number must be included but the Grashof number is omitted, being irrelevant in this case where buoyancy forces have relatively negligible effect.

The Reynolds number is a criterion of the type of fluid flow. At low velocities streamline flow (often termed laminar flow) occurs. At sufficiently high velocities mixed flow, usually termed turbulent flow is established. If the fluid is flowing through a tube, the size dimension l in the Reynolds number is the tube inside diameter d and the transition between laminar flow and turbulent flow takes place at a 'critical' Reynolds number of about 2300.

One widely accepted correlation for fluids being heated when in turbulent flow inside tubes is

$$\text{Nu} = 0{\cdot}023\ \text{Re}^{0\cdot8}\ \text{Pr}^{0\cdot4} \tag{24.6}$$

This correlation is found to be more accurate for gases than for liquids. If each side of equation (24.6) is divided by the product (Re Pr) then

$$\frac{\text{Nu}}{(\text{Re Pr})} = \frac{h}{\rho v c} = \text{St} = 0{\cdot}023\ \text{Re}^{-0\cdot2}\ \text{Pr}^{-0\cdot6} \tag{24.7}$$

St is the symbol for the Stanton number which is another useful dimensionless form of the heat transfer coefficient.

If the values of Pr are calculated from properties of gases listed in tables it will be observed that Pr varies little with temperature or pressure. A mean value of Pr equal to 0·72 may be assumed for air. Using this value, equations (24.6) and (24.7) may be simplified to give for air,

$$\text{Nu} = 0{\cdot}020\ \text{Re}^{0\cdot8} \tag{24.8}$$

$$\text{St} = 0{\cdot}028\ \text{Re}^{-0\cdot2} \tag{24.9}$$

Reynolds recognized that an analogy exists between convective heat transfer and fluid momentum transfer: both are due to the mixing caused by turbulence. If an element of fluid of mass m having a velocity v is reduced to zero velocity at the tube wall, the momentum transfer between the fluid and wall is mv and the heat transfer is $mc\theta$ where θ is the temperature difference. The ratio between the two is $c\theta/v$. The summation of the elemental momentum transfers per unit time per unit area perpendicular to the direction of heat transfer is the shear stress per unit area τ. The heat transfer rate $\dot{q}$ is then given by

$$\dot{q}/\tau = c\theta/v$$

Hence

$$\text{Nu} = \frac{\dot{q}d}{k\theta} = \frac{f}{2}\left(\frac{\rho v d}{\mu}\right)\left(\frac{\mu c}{k}\right) = \frac{f\,\text{Re Pr}}{2}$$

where f is the Fanning friction factor $\tau/\frac{1}{2}\rho v^2$.

$$\frac{f}{2} = \frac{\text{Nu}}{\text{Re Pr}} = \text{St} \tag{24.10}$$

This is a simple analogy based on the supposition that the flow is turbulent throughout and that there is no laminar boundary layer. If in the boundary layer the ratio of momentum diffusivity to thermal diffusivity (i.e. Pr) equals unity the result is still valid and this is approximately so for air.

The Fanning friction factor f is often considered to be a function only of the Reynolds number. For turbulent flow in clean round tubes, one empirical relationship used is

$$f = 0{\cdot}046\ \mathrm{Re}^{-0{\cdot}2} \tag{24.11}$$

Heat Transfer to Air by Forced Convection

The purpose of this experiment is to examine the empirical forced convection heat transfer and friction relationships by passing air through a heated tube. This can be achieved by making sufficient measurements to calculate the dimensionless numbers involved.

Experimental Details

Suitable apparatus is shown in Figure 24.1. Air is supplied by a blower, at a

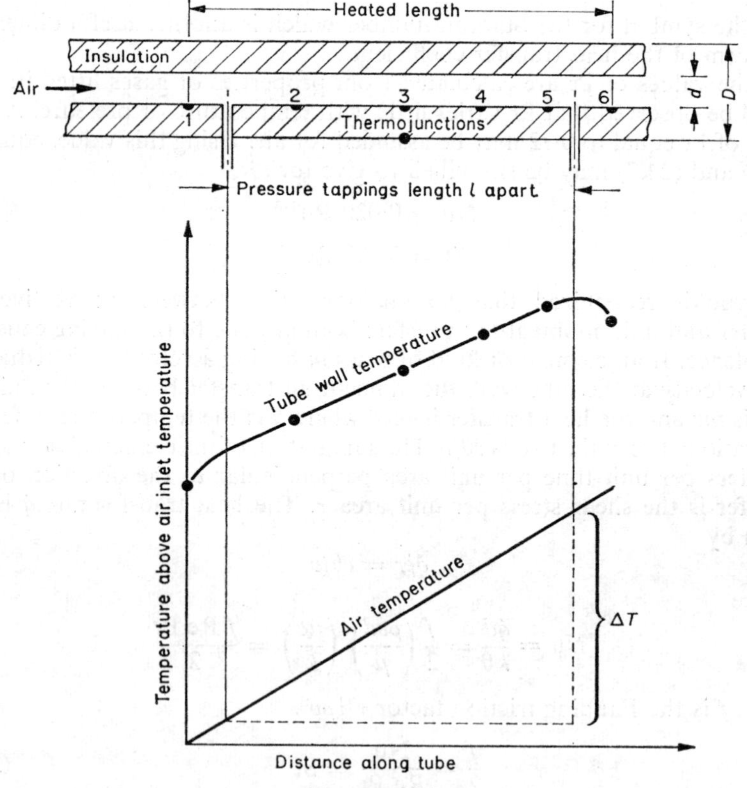

FIGURE 24.1

rate measured by a standard orifice (not shown) designed to British Standard Specification 1042. The inlet air temperature T_i is observed at the orifice plate in the supply pipe. The air is passed through a long metal tube. A downstream length of the tube is heated either by an electrically heated tape spiralled along its length or by a large alternating current passed along the tube, supplied at low voltage from a stepped transformer. The outside of the heated tube is thermally insulated from the atmosphere. The tube outer wall surface temperature at six stations along it and the air temperature at outlet are measured with thermocouples. If the cold junctions of the thermocouples are bound to the thermometer bulb near the orifice plate then temperatures are measured relative to the inlet air temperature, so giving temperature differences directly. One thermocouple is located on the inside and another on the outside of the insulation, for estimation of the radial heat transfer outwards at the centre section of the heated length. The pressure drop along a length of the tube is measured by an inclined water manometer connected across two pressure tappings, distance l apart.

Initially the blower inlet valve should be open fully, or if the air flow rate is controlled by blower speed, the blower should be set at maximum speed. This provides the maximum air mass flow rate, indicated by the greatest pressure drop across the orifice manometer. The electrical circuit may then be completed and the voltage increased. Some time is required before steady conditions prevail, indicated by the surface temperature recorded on thermocouple number 5 reaching a maximum. This maximum temperature must be limited in order to avoid damage. Tests are repeated at several different air flows: the reduction of the air flow in steps should be matched by decreasing the electrical power input to limit the maximum surface temperature. The minimum air flow rate used should still be large enough to give a Reynolds number greater than the 'critical' value, so that turbulent flow still occurs.

Along the central length of the heated tube the temperature gradient of the tube wall is linear, indicating an equal axial heat transfer by conduction into and out of this length, i.e., only radial heat transfer results from the electrical input at this station. The heat transfer rate (per unit area of heated tube surface) radially outwards by conduction through the insulation, $\dot{q}_r$, is found from the particular form of the Fourier equation, namely

$$\dot{q}_r = \frac{2k_r\,(\Delta T_r)}{\mathrm{d}(\ln D/d)} \tag{24.12}$$

where ΔT_r is the temperature difference measured across the insulation of known (or assumed) thermal conductivity, k_r. The heat transfer rate radially inwards (i.e. to the air flow) per unit area of heated tube surface, $\dot{q}$, is given by

$$\dot{q} = \frac{\text{electrical power input}}{\text{heated surface area}} - \dot{q}_r \tag{24.13}$$

There is a velocity and a temperature profile across any section of the heated length of tube so that the thermocouple at the exit station does not record accurately the air bulk temperature there. However the arithmetic mean of the air inlet temperature, T_i, and air outlet temperature given by this thermocouple is adequate to ascertain the mean specific heat of the air, c_m, from tables of properties of air. The air bulk temperature at outlet, T_0, is then evaluated from the energy balance

$$\dot{q} = \frac{\text{mass flow rate of air}}{\text{heated surface area}} \times c_m \times (T_i - T_0) \tag{24.14}$$

The arithmetic mean of T_i ($\doteqdot T_1$) and T_0 may be taken to be the air bulk temperature, T_m, at the station at the centre of the heated length. Properties of air at this temperature, T_m, which are required subsequently should be interpolated from tables of properties of air.

Dividing the air mass flow rate by the internal cross-section area of the tube gives the mass velocity, G, which by continuity must be equal to ρv. Hence the Reynolds number is given by Gd/μ.

The surface heat transfer coefficient, h, is obtained from the Newton equation

$$\dot{q} = h(T_1 - T_m) \tag{24.15}$$

and then the Nusselt number may be calculated from hd/k.

The Fanning friction factor, f, may be evaluated from

$$\Delta p = \frac{G^2}{g\rho}\left\{\frac{2fl}{d} + \frac{\Delta T}{T_m}\right\} \tag{24.16}$$

where ρ is the arithmetic mean air density, ΔT the air temperature increase (see Figure 24.1) and Δp the pressure decrease along the length, l, between the pressure tappings. Equation (24.16) is a simplified form of that which would be obtained by solving the momentum, continuity and energy equations for compressible flow with heat transfer. If the friction factor is determined with the heater switched off the last term of equation (24.16), which accounts for the momentum pressure loss due to acceleration by heating, disappears. The blower power, b.p. necessary to maintain the air circulation along the length, l, of the tube is given by

$$\text{b.p.} = \frac{\Delta p}{\rho} \times \text{mass flow rate of air} \tag{24.17}$$

At each measured air flow rate plot a graph of temperature along the tube wall as illustrated in Figure 24.1. The tube wall temperature at the centre station of the heated length can be read from the smooth curve obtained. The net heat transfer rate per unit area of heated tube surface, $\dot{q}$, is given by equation (24.13). After evaluating the mean specific heat of air, c_m, the air bulk temperature at outlet is calculated from equation (24.14). The air bulk temperature at the centre section is given by $\frac{1}{2}(T_i + T_o)$, on the assumption that the heat transfer rate to the air is uniform along the tube. The surface heat transfer coefficient is given by equation (24.15).

From this information, together with the dimensions of the apparatus and tables of properties of air, the Reynolds, Stanton and Nusselt numbers can be evaluated. From the corresponding pressure drops between the pressure tapping points, the Fanning friction factor can be determined using equation (24.16).

Discussion

Derive equation (24.5) (assuming $\beta g = 0$) by the method of indices. Discuss the assumptions made (for example, are the variables involved really 'independent' as stated in the analysis?)

Draw up a table for each experimental value of Reynolds number and enter (*a*) the experimental values of Nusselt and Stanton numbers and of the Fanning friction coefficient, and (*b*) the corresponding values obtained from equations (24.8), (24.9), (24.10) and (24.11). Calculate the percentage discrepancies between the values obtained in (*a*) and (*b*) and comment. Present the results by plotting $\log_{10}$ of Nusselt number, Stanton number, and Fanning friction factor against $\log_{10}$ of Reynolds number from the empirical relationships in equations (24.8), (24.9), (24.10), (24.11) and superimpose the experimental values.

Review the likely sources of error in the experiment (for example was it justifiable to ignore the tube wall thickness and to assume a value for the thermal conductivity of the insulation?).

The experimental observations reveal directly that the pressure drop, Δp, along the tube increases with mass flow rate, a result which could have been anticipated either intuitively or from equation (24.16). A prediction obtained by combining equation (24.16) and (24.17) is that the blower power for a particular mass flow rate is inversely proportional to the square of the density ρ^2. Thus if the coolant system were to be a closed loop, as in a gas-cooled reactor, it would be advantageous to operate at as high a pressure as practicable. It is apparent that for economy the ratio b.p./$\dot{q}$ should be low. For a given heat exchanger with a given coolant temperature rise and a given heat transfer rate (and hence a particular mass flow rate) deduce that

$$\text{b.p.}/\dot{q} \propto \frac{1}{c^3\rho^2}$$

Using tables of properties of gases show that, on this criterion for a coolant gas, carbon dioxide is about twice as attractive as air.

25

Heat Transfer by Conduction, Radiation, and Natural Convection

Heat transfer by natural convection results from fluid motion due to buoyancy forces, invoked by density changes resulting from temperature gradients in the fluid. The variables involved can be related by equation (24.4). The heat transfer coefficients are small compared with those obtained under forced convection conditions, and so accurate results are more difficult to obtain due, for example, to the appreciable effect of extraneous draughts. Further, the radiation heat transfer component can be a relatively large part of the total heat transfer rate, even with relatively low surface temperatures, and therefore may have to be accounted for. It is then necessary to know the emissivity of the surface so that radiation from it can be allowed for. The emissivity is the ratio of the rate of thermal emission from that surface to the rate of emission at the same temperature from a surface (termed a 'black body') which would absorb all thermal radiation incident upon it. The component of the heat transfer by radiation is given by the Stefan-Boltzmann equation

$$\dot{q}' = \varepsilon\sigma(T_o^4 - T_a^4) \tag{25.1}$$

where $\dot{q}'$ is the radiation heat transfer rate per unit area;
ε is the surface emissivity, assumed constant;
σ is the Stefan-Boltzmann constant, in appropriate units;
T_0 is the surface absolute temperature;
and T_a is the bulk fluid absolute temperature.

Experimental Details

A horizontal metal tube containing a long electrical heater element is encased in a thick layer of plaster (Figure 25.1). The ratio of the cylinder length to the outside diameter of the plaster is not less than 3 and the ends insulated so that axial heat transfers may be neglected. The cylinder is suspended horizontally in a draught free atmosphere. Three thermojunctions (*a*, *b* and *c*) are spaced at the mid-section of the cylinder between the outside of the metal tube and the plaster. After steady conditions have been achieved the three

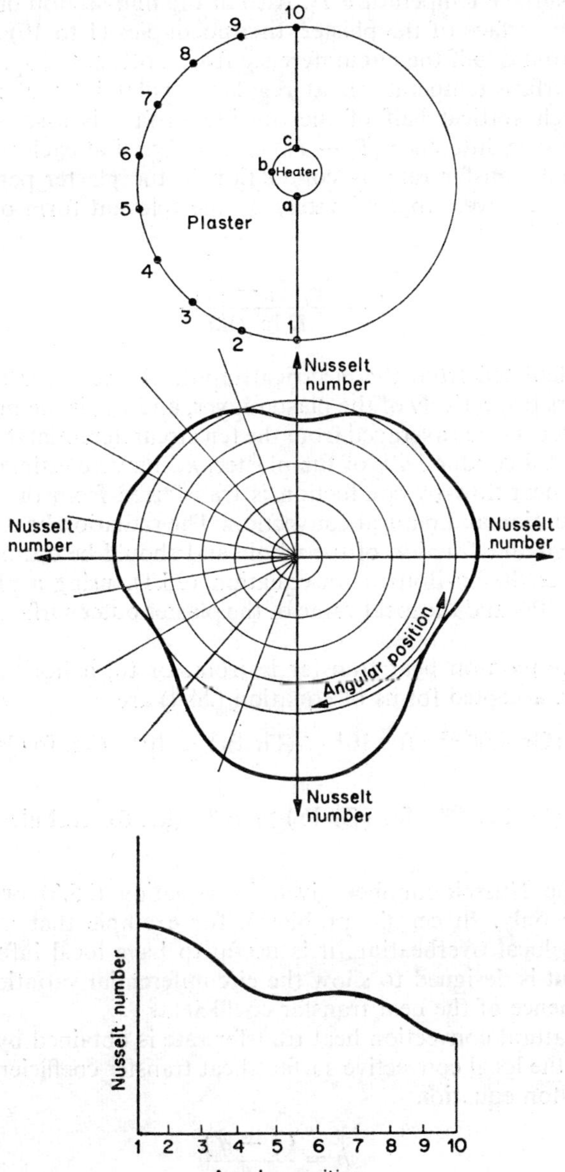

1
2
3
4
5
6
7
8
9
10
a
b
c
Heater
Plaster
Nusselt number
Nusselt number
Nusselt number
Nusselt number
Angular position
Nusselt number
1 2 3 4 5 6 7 8 9 10
Angular position

FIGURE 25.1

thermocouples record almost the same temperature which is taken to be the plaster inner surface temperature T_i. Also at the mid-section of the cylinder, on the outside surface of the plaster, thermocouples (1 to 10) are spaced at 20° intervals round half the circumference from bottom to top. These record the outside surface temperatures at regular angular intervals: symmetrical cooling of each vertical half of the outside surface is assumed. Thus the radial temperature difference $(T_i - T_0)$ can be found at each sector interval. The radial heat transfer rate by conduction in the plaster per unit area of outside surface is given approximately by the relevant form of the Fourier equation

$$\dot{q} = \frac{2k_r(T_i - T_{0m})}{D \ln D/d} \tag{25.2}$$

where $\dot{q}$ is calculated from the electrical input, D and d are the outer and inner diameters respectively of the plaster layer, and T_{om} is the mean temperature of the outer surface averaged from the ten circumferential thermocouples. Thus the thermal conductivity of the plaster, k_r, may be estimated.

This radial heat flow by conduction is transferred from the outer plaster surface by radiation and natural convection. The radiation heat transfer rate, $\dot{q}'$, is not a negligible fraction of the whole and should be estimated for each sector from the Stefan-Boltzmann equation (25.1), using a plaster surface emissivity, $\varepsilon = 0{\cdot}9$ and the local value of the plaster outer surface temperature T_o.

If natural convection heat transfer is from, or to, a horizontal cylinder exposed in air, accepted forms of equation (24.4) are

$$\text{Nu} = 0{\cdot}52\,(\text{Gr Pr})^{0{\cdot}25} \quad \text{for } 10^4 < (\text{Gr Pr}) < 10^9 \quad \text{(i.e. for laminar flow)} \tag{25.3}$$

$$\text{Nu} = 0{\cdot}10\,(\text{Gr Pr})^{0{\cdot}33} \quad \text{for } (\text{Gr Pr}) > 10^9 \quad \text{(i.e. for turbulent flow)} \tag{25.4}$$

However the Nusselt number given by equation (25.3) or (25.4) is an average value only. In certain problems, for example that of boiler tube failure due to local overheating, it is useful to have local information and this experiment is designed to show the circumferential variation of Nusselt number and hence of the heat transfer coefficient.

The local natural convection heat transfer rate is obtained by subtraction, $(\dot{q} - \dot{q}')$, and the local convective surface heat transfer coefficient determined from the Newton equation

$$h = \frac{(\dot{q} - \dot{q}')}{\theta} \tag{25.5}$$

where θ is the local temperature difference between the plaster outer surface, T_o, and the air bulk temperature, T_a. The local Nusselt number may then be determined. The Grashof number and the Prandtl number can be evaluated also: the variation of Grashof number and Prandtl number is small and the value of each may be averaged. As usual, the properties of air are taken from tables at the film temperature, which is assumed to be $\frac{1}{2}(T_o + T_a)$. Plot the Nusselt number against angular position round the tube, both by cartesian coordinates and polar coordinates. Also obtain an average value of the Nusselt number.

Discussion

Explain the shape of the curves obtained by considering the flow pattern over the tube. Compare the average Nusselt number obtained with that predicted by the appropriate equation—either (25.3) or (25.4). For large bodies, free from cavities, these equations are often approximated with sufficient accuracy by

$$q = \text{constant } \theta^{1 \cdot 25}$$

Consult a standard text-book on heat transfer for the value of the constant, which depends on the units used, and compare the measured electrical energy input with the heat transfer predicted by this equation using $\theta = T_{om} - T_a$.

Draw up a relaxation plot of the isothermal and adiabatic lines through the plaster, taking the measured temperatures at the inner and outer plaster surfaces as boundary conditions, to illustrate pictorially the error in the assumption inherent in equation (25.2) that heat is transferred radially and uniformly outwards.

Discuss the likely source of errors in the experiment.

One thermocouple on the inside and one on the outside of the plaster would suffice if provision was made to rotate the tube through the 20° intervals. Consider the advantages and disadvantages of this alternative experimental arrangement.

26

Heat Transfer by combined Conduction and Convection

When a rib or fin projects from a wall into a fluid at a different temperature, heat transfer occurs through the fin by conduction and at the fin surface mainly by convection. These heat transfers take place concurrently. Examples of this simultaneous conduction and convection occur for (*a*) the fins on the cylinder barrel of an internal combustion engine, when the heat transfer is from the barrel to the air via the surface of the fins and (*b*) a gas turbine blade, clamped at its root to the relatively cool turbine rotor, when the heat transfer is from the high temperature gas via the surface of the turbine blade to the rotor.

A situation similar to that illustrated by (*a*) may be studied both analytically and experimentally by considering the heat transfer rate to the surrounding air from a long bar, with constant cross-section area, heated at one end for sufficient time until the temperature distribution along it is steady. Either an electrical heater, thermostatically controlled, or a steam jacket in which the pressure is controlled, may be used at the heated end. The temperatures, at known distances along the bar, are measured by thermometers inserted in drilled holes or by thermojunctions attached to the surface of the bar. Hence, providing the atmospheric temperature has been observed, Table 26.1 can be completed.

TABLE 26.1 Observations of temperature along a bar heated at one end.

Distance from heated end, x	$x_0 = 0$	x_1	x_2	x_3	x_4	x_5
Temperature above atmosphere, θ	$\theta_o =$					
θ_o/θ	1					
$\ln \theta_o/\theta$	0					

Theory

To obtain a simple analytical solution, heat transfers by radiation from the bar will be neglected; alternatively the radiation component may be considered to be included in the convection effect. Consider, with reference to Figure 26.1, the energy balance for an elemental length dx, of the bar. If steady state conditions exist then the heat conduction $\dot{q}_x$ along the bar

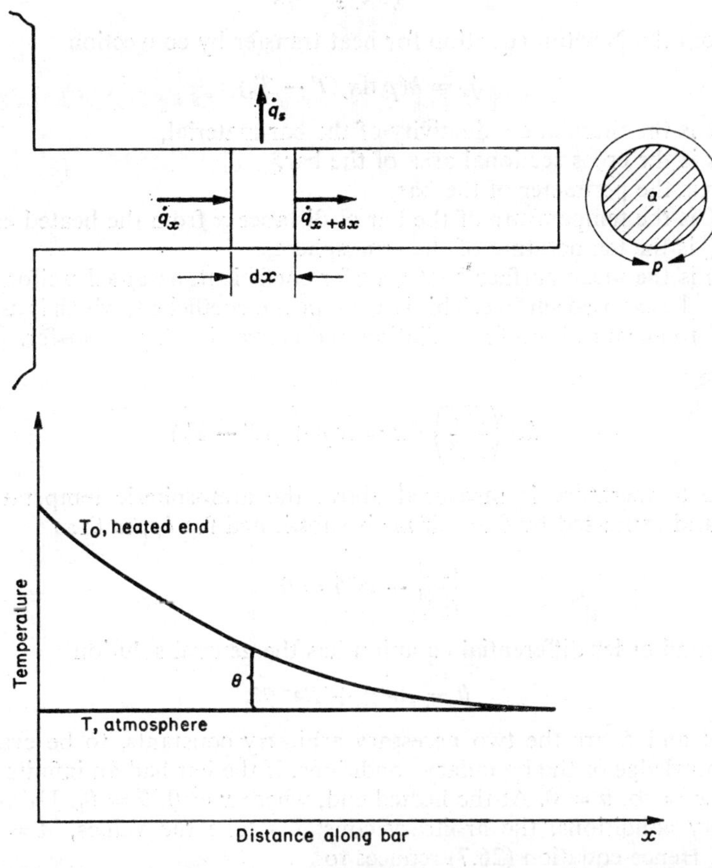

FIGURE 26.1

towards the element will equal the sum of the heat conduction from the element and the heat transfer to the air at the surface of the element, $\dot{q}_s$,

$$\dot{q}_x = \dot{q}_{x+\mathrm{d}x} + \dot{q}_s \tag{26.1}$$

Assuming that the temperature is uniform over a plane cross-section, then, from the differential form of the Fourier equation for heat transfer by conduction

$$\dot{q}_x = -ka\left(\frac{\mathrm{d}T}{\mathrm{d}x}\right)_x \tag{26.2}$$

$$\dot{q}_{x+dx} = -ka\left(\frac{dT}{dx}\right)_{x+dx}$$

$$= -ka\left(\frac{dT}{dx}\right)_x + \frac{d^2T}{dx^2}\,dx \tag{26.3}$$

Also from the Newton equation for heat transfer by convection

$$\dot{q}_s = h(p\,dx)(T - T_a) \tag{26.4}$$

where k is the thermal conductivity of the bar material,
a is the cross sectional area of the bar,
p is the perimeter of the bar,
T is the temperature of the bar at distance x from the heated end,
T_a is the temperature of the atmosphere,
and h is the mean surface heat transfer coefficient, assumed uniform over the exposed surface. This is a composite coefficient, which is assumed to account both for radiation and convection heat transfer.

Hence

$$ka\left(\frac{d^2T}{dx^2}\right)dx = h(p\,dx)(T - T_a) \tag{26.5}$$

If the temperature is measured above the atmospheric temperature as datum and expressed by θ and if m^2 is substituted for hp/ka then

$$\frac{d^2\theta}{dx^2} - m^2\theta = 0 \tag{26.6}$$

This second order differential equation has the general solution

$$\theta = Ae^{mx} + Be^{-mx} \tag{26.7}$$

where A and B are the two necessary arbitrary constants, to be evaluated from knowledge of the boundary conditions. If the bar had an infinite length then at $x = \infty$, $\theta = 0$. At the heated end, where $x = 0$, $\theta = \theta_o$. Using these boundary conditions, the arbitrary constants take the values, $A = 0$ and $B = \theta_o$. Hence equation (26.7) reduces to

$$\theta = \theta_o e^{-mx} \tag{26.8}$$

which relates the excess temperature, θ above atmospheric with the distance x along the bar.

The heat transfer rate from the infinitely long bar is

$$\int_{x=0}^{x=\infty} \dot{q}_s\,dx$$

Substituting for $\dot{q}_s$ from equation (26.4) and then for θ from equation (26.8) and integrating between the limits yields

$$\dot{q}_{0\to\infty} = kam\theta_o \tag{26.9}$$

Equation (26.8) may be written in the form

$$\ln \frac{\theta_o}{\theta} = mx \tag{26.10}$$

Experimental Details

Observe the temperatures at measured distances along the bar (Table 26.1), and plot ln (θ_o/θ) against distance x. This should yield a straight line through the origin with a gradient m. From the values of $m = (hp/ka)^{\frac{1}{2}}$ estimate the mean surface heat transfer coefficient h. The bar dimensions are known and the thermal conductivity of the material of the bar can be taken from tables of thermal properties. Hence the rate of heat loss from the bar, $\dot{q}_{0=\infty}$, may be evaluated.

Discussion

Deduce equation (26.8) from first principles and include a full statement of the assumptions which must be made. Extend the analysis to the case of a bar of length L and, by assuming that at $x = L$ the temperature gradient is zero (i.e. $d\theta/dx = 0$), show that the temperature distribution along such a finite bar is given by

$$\theta = \theta_o \left[\frac{\cosh m(L - x)}{\cosh mL}\right] \tag{26.11}$$

Hence show that the heat transfer from the finite bar is

$$\dot{q}_{0 \to L} = kam\theta_o \tanh (mL) \tag{26.12}$$

For the bar used in the experiment determine the percentage error in the heat transfer rate from the bar due to the assumption that it was infinitely long.

Two very long bars of different metals have the same dimensions and the same mean surface heat transfer coefficients. If the thermal conductivity of one metal is known, outline an experiment and the appropriate analysis which would permit the thermal conductivity of the other metal to be determined.

27

Heat Transfer with Change of Phase

In many processes a transfer of heat is accompanied simultaneously by a transfer of mass and there are analogous features between these two types of transfer. One situation in which mass transfer occurs in conjunction with convective heat transfer will be considered here.

The conversion of vapour into liquid (i.e. condensation), takes place at constant temperature, termed the saturation temperature, which is a function of the fluid pressure. The process is accompanied by the release of the energy of evaporation. The magnitude of the energy released is large, particularly with a vapour such as steam. In a steam condenser this energy is transferred as heat through the walls of tubes to cooling water flowing within the tubes. During the process there is a transfer of mass from the vapour space to the condensate deposited as liquid on the outside surface of the tubes. On the cooled tube walls the condensate forms in either a 'filmwise' or a 'dropwise' manner. A high 'steam side' heat transfer coefficient can be obtained with dropwise condensation but it is difficult to provide conditions in which this will endure. Filmwise condensation usually predominates, resulting in a lower heat transfer coefficient due to the thermal resistance caused by the insulating effect of the liquid film on the tube surface. However, this experiment demonstrates that the main resistance to heat transfer is not on the steam side of the tube wall, whatever the manner of condensation, but on the cooling water side.

Figure 27.1 shows an outline of a single tube surface condenser. Saturated steam is supplied to the jacket of the condenser. Cooling water is circulated through the tube of known dimensions, and the mass flow rate of water is metered. The inlet and outlet cooling water temperatures are measured: the steam temperature is taken to be the saturation temperature corresponding to the measured condenser pressure.

The purpose of the experiment is to demonstrate that the magnitude of the heat transfer rate from the condensing steam to the cooling water depends upon (*a*) the velocity of the cooling water within the tube, and (*b*) the steam pressure in the condenser.

For (*a*) the condenser is operated with constant pressure over a range of recorded water flow rates which are within the turbulent flow region (i.e.

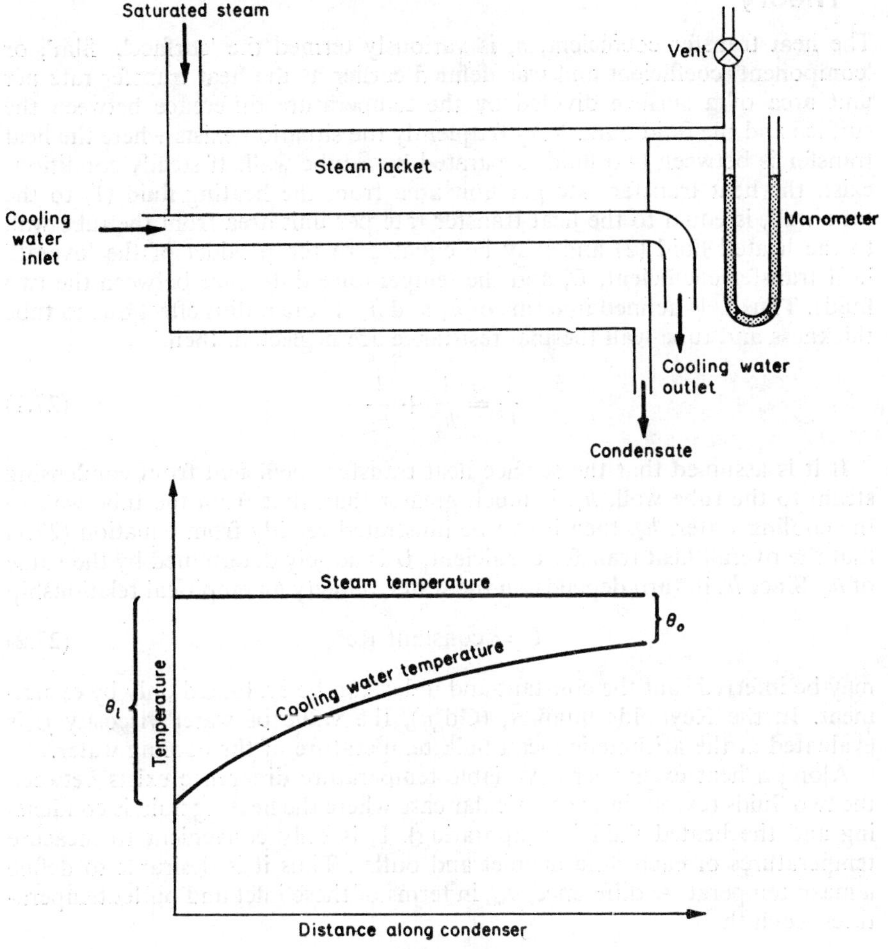

FIGURE 27.1

Re > 2300). The water flow rate should be approximately doubled at each step to give a suitable spread of experimental points, and the cooling water temperatures recorded.

For (*b*) the condenser is operated with a constant water flow rate over a range of condenser pressures. The steam pressure should be reduced in stages from atmospheric to the maximum vacuum obtainable: again the water temperatures and the (constant) mass flow rate of cooling water are recorded.

After each change in experimental conditions, adequate time must be allowed for steady conditions to be re-established before observations are made.

Theory

The heat transfer coefficient, h, is variously termed the 'surface', 'film', or 'component' coefficient and was defined earlier as the heat transfer rate per unit area of a surface divided by the temperature difference between the surface and the fluid bulk. Very frequently the situation exists where the heat transfer is between two fluids separated by a tube wall. If steady conditions exist, the heat transfer rate per unit area from the heating fluid (1) to the tube wall, is equal to the heat transfer rate per unit area from the tube wall to the heated fluid (2) and may be equated to the product of the 'overall' heat transfer coefficient, U, and the temperature difference between the two fluids. Thus U is defined in terms of h_1 and h_2. If the radius effect due to tube thickness and tube wall thermal resistance are neglected, then

$$\frac{1}{U} = \frac{1}{h_1} + \frac{1}{h_2} \tag{27.1}$$

If it is assumed that the surface heat transfer coefficient from condensing steam to the tube wall, h_1, is much greater than that from the tube wall to the cooling water, h_2, then it can be illustrated readily from equation (27.1) that the overall heat transfer coefficient, U is largely determined by the value of h_2. Since h_2 in turn depends on the water velocity an empirical relationship

$$U = \text{constant Re}^x \tag{27.2}$$

may be inferred, but the constant and index can be evaluated only by experiment. In the Reynolds number, $(\text{G}d/\mu)$, the value of water viscosity μ is evaluated at the arithmetic mean bulk temperature of the cooling water.

Along a heat exchanger a variable temperature difference exists between the two fluids (except in the particular case where the heating fluid is condensing and the heated fluid is evaporating). It is only convenient to measure temperatures of each fluid at inlet and outlet. Thus it is desirable to define a mean temperature difference, θ_m, in terms of these inlet and outlet temperatures, such that

$$\theta_m = \dot{q}/U \tag{27.3}$$

where $\dot{q}$ is the heat transfer rate per unit area. By equating the heat transfer rate from the heating fluid to the heat transfer rate to the heated fluid it can be deduced that

$$\theta_m = \frac{\theta_i - \theta_o}{\log \theta_i/\theta_o} \tag{27.4}$$

where θ_i and θ_o are the temperature differences between the fluids at the inlet and outlet of the heat exchanger respectively. The term θ_m is usually referred to as the 'logarithmic mean temperature difference'. This relationship, ascribed to Grashof, is valid for fluids in counter flow, parallel flow, or cross flow providing that the specific heats and surface heat transfer coefficients of the fluids are almost constants.

Experimental Details

From the tube dimensions, tabulated data on water viscosity, and the observations for part (*a*) of the experiment, evaluate the overall heat transfer coefficients U and the corresponding Reynolds numbers. Plot ln U *versus* ln Re; the gradient of the resulting smoothed straight line gives the value of the index x in equation (27.2): consequently the constant may be evaluated. From observations obtained under part (*b*) of the experiment plot a graph of heat transfer rate against absolute steam pressure.

Discussion

Select a value of Reynolds number and obtain the corresponding value of overall heat transfer coefficient from the completed version of equation (27.2). Use equation (24.6) to estimate the surface heat transfer coefficient on the water side, h_2, and hence from equation (27.1) evaluate the surface heat transfer coefficient on the steam side, h_1 (usually termed the condensation coefficient).

The report should present a deduction of equation (27.1) including the effect of tube wall thickness, and equation (27.2).

Discuss the usefulness and the limitations of the results obtained in the experiment. Indicate the most likely causes of scatter of the experimental results. Explain why the graphs obtained under (*a*) and (*b*) would only approximate to straight lines even if this scatter could be avoided. Indicate which experimental results in part (*b*) are most likely to be in error due to the neglect of (i) tube wall thermal resistance and (ii) the presence of some air in the condenser.

28

Heat Transfer Through Vacua

Vacua are used in many pure and applied sciences [1]. They are of interest not primarily for their own sake but because they provide a means of achieving suitable working conditions, e.g. clean atmospheres when growing crystals, or the removal of molecules that would otherwise scatter high speed electrons in a T.V. tube.

The subjects of vacuum technology and cryoengineering are also inextricably linked because thermal insulation, being a prime requisite in cryoengineering is frequently obtained by the use of vacuum (e.g. as in the envelope of a dewar flask). For these reasons, the subject of vacuum technology will be introduced into this course as a 'means' and not as an 'end-in-itself'.

Gases have thermal conductivities so much lower than solids, that they are often used for thermal insulation—hence the cavities in double glazing and furnace walls. Nevertheless appreciable heat leaks occur across such gas interspaces. Thus the purpose of the present experiment is to study the thermal flux between two bodies at different temperatures, so that the components due to convection, gaseous conduction, radiation and solid conduction can be measured accurately and interpreted. Such a quantitative analysis is often a desirable prerequisite to the design of a system through which it is required to either inhibit or enhance the rates of heat flow across a gas space. The analysis is achieved by the use of gas pressure as a variable, and may be applied profitably to many systems, even some which in practice will be employed entirely at atmospheric pressure (e.g. a domestic oven).

In the particular experimental rig considered here the heated body is completely surrounded by another body (e.g. as in Figure 28.1) but such a geometrical requirement is not necessary for the present technique to be adopted.

A typical form of the variation of heat leak with gaseous pressure is given in curve 1 of Figure 28.2, for which T equals T_1. The particular shape of the curve (e.g. the pressures corresponding to the onset, B_1, of gaseous conduction and the commencement, F_1, of convection) depends upon the geometrical configuration, the temperatures involved and the gas present. For example if the rate of heat dissipation in the enclosed body is increased so that its absolute temperature rises from T_1 to T_2, then the steady state heat leak will now be described by a curve such as 2. If on the other hand the original gas is replaced by one with a higher thermal conductivity, the

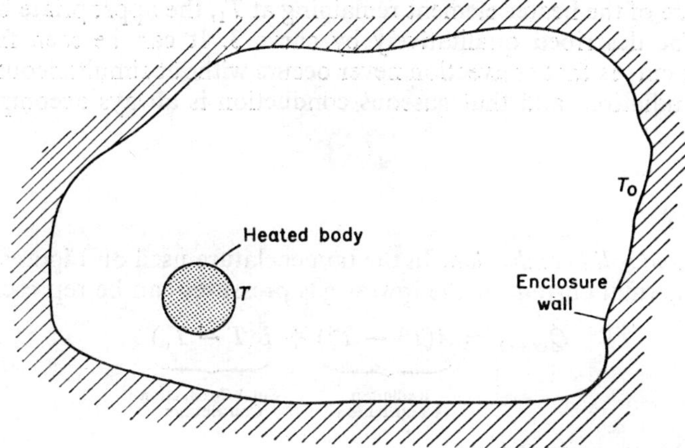

FIGURE 28.1

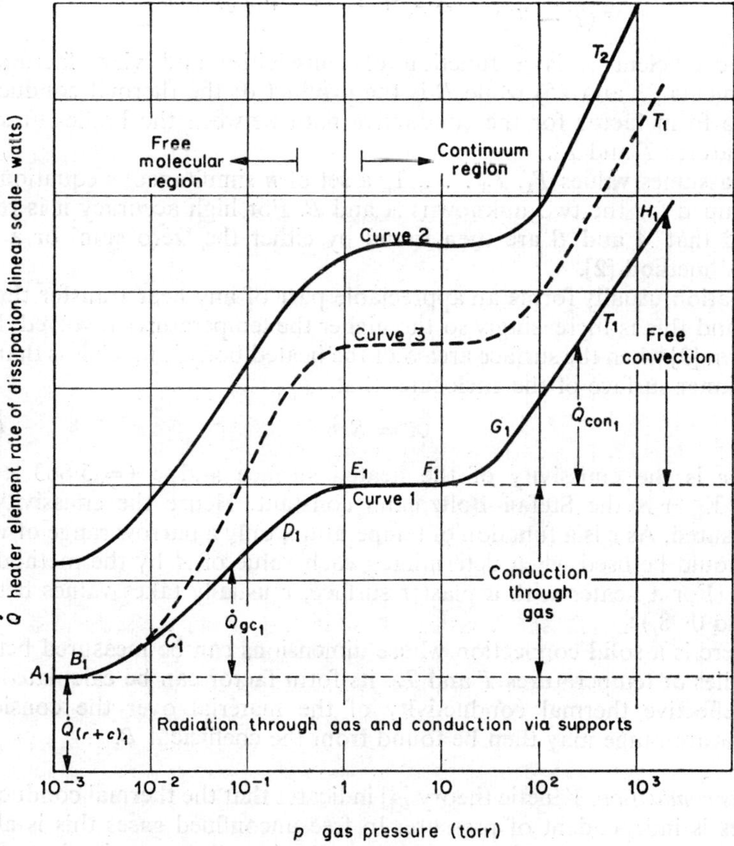

FIGURE 28.2

temperature of the heater element remaining at T_1, the appropriate behaviour will now be described qualitatively by curve 3. It can be seen from such families of curves that convection never occurs without simultaneous conduction and radiation, and that gaseous conduction is always accompanied by radiation.

Theory

Radiation and solid conduction. In the nomenclature used on Figures 28.1 and 28.2, the thermal current at the lowest gas pressures can be represented by

$$\dot{Q}_{(r+c)} = \underbrace{A(T^4 - T_0^4)}_{\text{Radiation Contribution}} + \underbrace{B(T - T_0)}_{\text{Solid Conduction Contribution}} \tag{28.1}$$

This may be rearranged as

$$\frac{\dot{Q}_{(r+c)}}{(T - T_0)} = A(T + T_0)(T^2 + T_0^2) + B \tag{28.2}$$

The coefficient A is a function of emissivities and view factors (see Experiments 23 and 25) while B is the product of the thermal conductivity and the form factor for the conductive path between the bodies at steady temperatures T and T_0.

If T assumes values $T_1, T_2 \ldots, T_n$ a set of n simultaneous equations can be obtained for the two unknowns A and B. For high accuracy it is recommended that A and B are then found by either the 'zero sum' or a 'least squares' method [2].

Radiation usually forms an appreciable part of any heat transfer through a gas, and this is increasingly so the higher the temperatures involved. It can be shown [3] when the surface area S of the heated body is much less than that of the inner surface of the enclosure, that

$$A = S\varepsilon\sigma \tag{28.3}$$

where ε is the emissivity of the heated surface and σ ($= 5{\cdot}663 \times 10^{-8}$ W m^{-2} K^{-4}) is the Stefan–Boltzmann constant. Hence the emissivity can be measured. As ε is a function of temperature, only a narrow range of values of T should be used when determining each value of A by the method described. (For a heater with a plaster surface, ε usually takes values between 0·95 and 0·98.)

If there is a solid connection whose dimensions can be measured between the bodies at temperatures T and T_0, its form factor can be calculated. The mean effective thermal conductivity of the material over the considered temperature range may then be found from the coefficient B.

Gaseous conduction. Kinetic theory [4] indicates that the thermal conductivity of gases is independent of pressure. In free unconfined gases this is always so because the number of molecules contributing to the conduction of heat increases, while their mean free path (i.e. the average distance a molecule travels between successive collisions with other gas molecules) decreases resulting in more energy scattering, both in proportion to the pressure. These two processes result for example in the plateau E_1F_1 shown in Figure 28.2.

At pressures below that corresponding to E_1, the molecular mean free path becomes greater than the heater-to-wall separation in the containing vessel. Hence, in this region, the molecules are more likely to hit the vessel walls or the heater surface rather than collide with one another. For gaseous conduction well within this 'free-molecular region' it can be shown [5] that

$$Q_{gc} = \frac{(\gamma + 1)}{4(\gamma - 1)} \sqrt{\left[\frac{2R}{\pi T_0 M}\right]} \alpha (T - T_0')p \qquad (28.4)$$

if (i) the gas is monatomic (e.g. helium), and (ii) the rate of energy dissipation in the heater does not appreciably affect the temperature of the whole environment. In this equation p represents the gas pressure; $\alpha(=(T_r - T_i)/(T_s - T_i))$ is the accommodation coefficient for gas molecules accepting energy from the surface; T_g, T_i and T_r are the temperatures of the surface, the incident and the reflected molecules respectively; R, M and γ are the universal gas constant, molecular weight and the ratio of the principal specific heats respectively.

TABLE 1

Typical values for the parameters M, γ and α that appear in equation (28.4)

Gas	He	Ne	Ar
Molecular Weight M	4·003	20·183	39·944
Ratio of Principal Specific Heats γ	1·66	1·64	1·67
Typical values of Accommodation Coefficient α	0·50	0·74	0·80

The temperature T_0' appearing in equation (28.4) is a function of the mean free path of the molecules and hence of the pressure. If however the mean free path considerably exceeds the heater-to-wall distance then T_0' becomes equal to T_0. By plotting $\dot{Q}_{gc}$ versus p on linear-linear graph paper, a straight line through the origin results for the free molecular region with T and T_0 fixed. Assuming that T_0' equals T_0, it is possible from the gradient of this line to evaluate α. (An extensive compilation of experimentally determined values for α appears in reference 5).

If measurements are made with a succession of different gases in the enclosure (all other conditions remaining identical) their thermal conductivities will be in proportion to their respective $\dot{Q}$ increments between the A_1B_1 and E_1F_1 plateaux.

The basic principles of the Pirani (or hot-wire) vacuum gauge are that the electrical resistance of a wire depends upon its temperature, which in turn depends upon the rate of loss of heat from the filament. This is a function of the number (and type) of molecules hitting the wire per second and so of the gas pressure. Hence by observing the variation of the electrical resistance of the wire the appropriate changes of pressure may be indicated. The heat transfer analysis for this gauge is, as can be seen from this outline analysis, very similar to that for the present experimental rig. Hence can you explain why the Pirani gauge is mostly used for the pressure range from 1 to 10^{-3} torr?

Free convection. This occurs in a constant pressure fluid due to buoyancy forces produced by density changes resulting from the existence of temperature gradients. The rates of heat transfer obtained by free convection are usually small compared with those obtained under forced convective conditions.

From the experimental measurements it may be deduced [6] that the convective heat flux

$$\dot{Q}_{\text{con}} = C(T - T_0)^{1 \cdot 25} \qquad (28.5)$$

The constant C can be determined from the log-log plot of the data.

In many cases in science, it is possible to unify, and thereby simplify the presentation of, a range of experimental results (e.g., for different geometries, temperature distributions and gases) by employing the data in non-dimensional forms. This is especially true for gaseous heat leaks, which can be described in terms of three dimensionless parameters:

(i) *Nusselt number*, (Nu), which for this experiment is defined as the ratio of the sum of the convective plus gaseous conductive heat transfers per unit surface area to what would be the continuum conductive heat transfer under the prevailing temperature conditions.

$$\text{Nu} = hd/k$$

(ii) *Grashof number*, (Gr), which is defined as the ratio of the buoyancy force per unit area to the viscous drag force per unit area.

$$\text{Gr} = \left\{\frac{\beta g(T - T_0)\, d^3\rho^2}{\mu^2}\right\}$$

(iii) *Prandtl number*, (Pr), which is defined as the ratio of the kinematic viscosity to the thermal diffusivity of the fluid.

$$\text{Pr} = \left(\frac{\mu C_p}{k}\right).$$

Here C_p represents the specific heat of the gas at constant pressure,
d represents the diameter of the heater element,
g represents the acceleration due to gravity,

h represents the heat transfer coefficient for combined convection and gaseous conduction, i.e. the rate of energy flow per unit area per unit temperature difference,
k represents the thermal conductivity of the gas at temperature $\frac{1}{2}(T_0 + T)$,
β represents the coefficient of volume expansion of the gas at constant pressure ($\simeq 2/(T_0 + T)$ for a perfect gas at temperature $\frac{1}{2}(T_0 + T)$),
ρ and μ represent the density and viscosity of the gas respectively at temperature $\frac{1}{2}(T_0 + T)$.

For the considered geometry, the onset of convection (i.e., the F_1G_1 knee in Figure 28.2) is generally accepted to occur at a Gr of approximately 2400. Does this agree with your measurements?

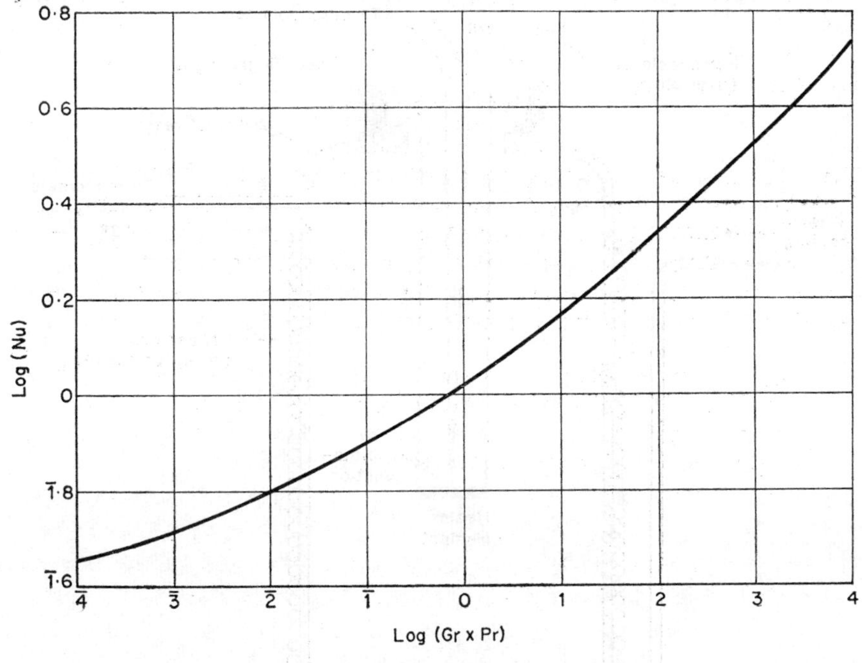

FIGURE 28.3

Plotting $\log_{10}$ (Nu) versus $\log_{10}$ (Gr.Pr) should reveal that within experimental error a single curve is obtained of the form shown in Figure 28.3. If the data can be approximately described by

$$\mathrm{Nu} = D(\mathrm{Gr.Pr})^m \qquad (28.6)$$

determine the values of the constant D and the exponent m for the range of Rayleigh numbers (i.e. products of Gr and Pr) covered in your particular experiment. From this procedure you will come to appreciate some of the power of the non-dimensional approach, which enables data from relatively few experiments to be used for interpreting a wide range of related events.

Experimental Details

Test Equipment. A cylindrical heater element is suspended horizontally from a top cover-plate of a steel vessel, which is capable of withstanding an internal pressure of 10 000 torr. The element consists of a plaster-covered tube, within which is a heater coil that can be supplied with electrical energy at a measured rate from outside the vessel. The electrical energy dissipated in the heater is mainly transferred through the gas, but some is conducted to the top cover plate via the supports and electrical leads. These rates of

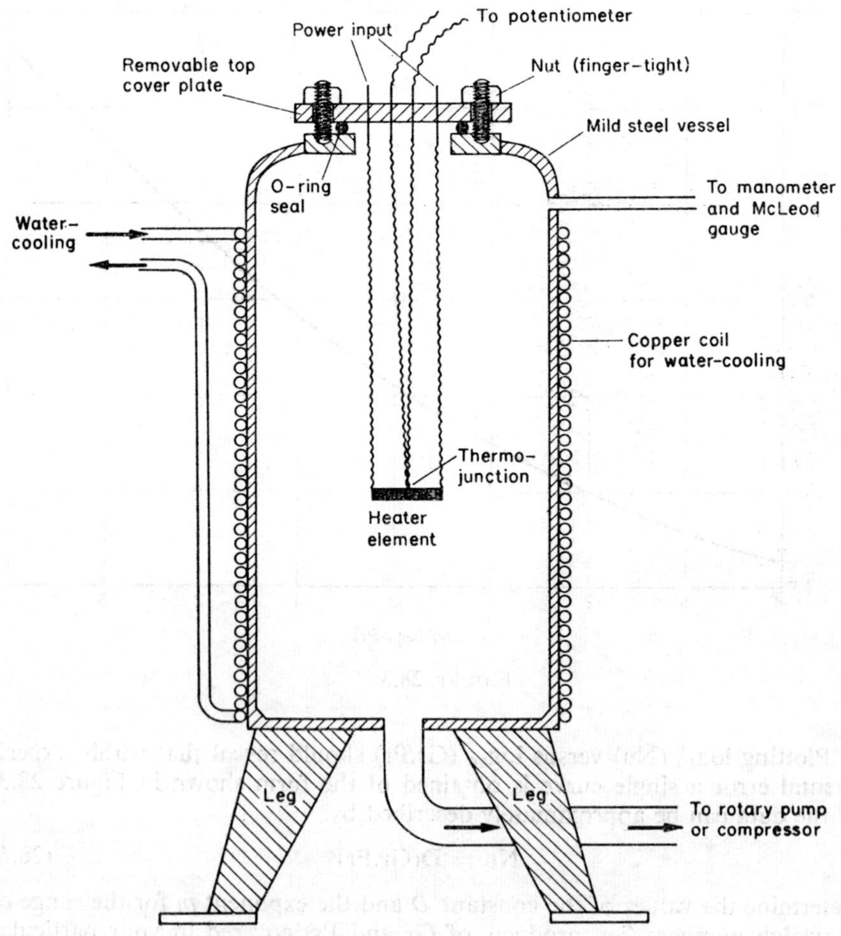

FIGURE 28.4

conduction can be calculated and remain, within experimental error, constant (irrespective of gas pressure) for the heater element at a given T. Thus the curve A_1 B_1 C_1 D_1 E_1 F_1 G_1 H_1 of Figure 28.2 would merely have its ordinates reduced by this fixed rate of solid conduction, in order for it to represent the rate of heat transfer through the gas alone. The heater element is sufficiently small and remote from the walls of the large vessel, to ensure substantially free convection. All the electrical leads pass through, but are electrically insulated from the demountable cover-plate, which is sealed in position on the vessel by an O-ring.

Pressures above and below atmospheric can be obtained with a combined compressor and vacuum pump connected by a short, wide pipeline to the vessel. It may be necessary to use a vapour pump as well as a rotary pump to attain sufficiently high vacua to eliminate gaseous conduction in the vessel. This however depends upon the particular experimental arrangement employed, and whether a single- or a two-stage rotary pump is used as well as upon the flow resistance of the pumping lines. Remember silicone fluid vapour pumps usually only function when their 'backing pressures' are less than 10^{-1} torr.

Any rotating machinery (e.g. the rotary pump) in the vicinity of the experimental rig should be placed upon antivibration mounts. Also the rotary pump must be connected to the pumping line by a flexible hose (see Volume I, Chapter 3). If these rules are not obeyed, vibration can be transmitted to the heater element and so spuriously high rates of heat transfer occur.

The vessel may be charged with other clean, dry gases (e.g. He or CO_2) from high pressure cylinders. Appropriate connections, control valves, and safety blow-offs must be included so that the vessel and lines are not subjected to explosive pressures. When it is proposed to use a gas other than air, the system should be evacuated, leak tested with a Pirani or mass spectrometer leak detector, and if leak tight, flushed-out with the type of gas that is to be used. The system is evacuated once again and then filled with the gas to the maximum pressure at which measurements are to be made. The experiment should then be completed as quickly as possible to minimize the contamination of the gas by air desorbed from the walls.

Appropriate pressure gauges, connected by suitable valves and short non-flow lines, indicate the gas pressure within the vessel. For example, a manometer with the mercury surface in one limb exposed to atmospheric pressure and the unknown overpressure on the other mercury surface could be used to show absolute pressures up to 2000 torr. (This requires the atmospheric pressure to be measured simultaneously with a Fortin barometer.) For pressures down to 30 torr, the mercury manometer could again be used but this time the other limb would be permanently evacuated (and sealed via a glass tap) to a pressure of less than 10^{-2} torr. To measure pressures lower than 1 torr a McLeod gauge (which is based on Boyle's law) is adopted. The lowest pressure (of non-condensible gases) which can be measured accurately with this instrument is limited by the vapour pressure of mercury ($\sim 2 \times 10^{-3}$ torr) at room temperature. For lower pressures in the present system a Pirani gauge is used.

If sufficiently low pressures are not attainable in the vessel then it is likely that either there is an air leak, which must be repaired before proceeding, or contamination of the vacuum system has occurred. The latter may require that the 'dirty' oil in the rotary pump has to be changed, the contaminant usually being a high vapour pressure liquid such as water (or methylated spirits, which is often used for leak location).

The outside surface temperature at mid-length of the plaster covered tube is measured with a chromel-alumel thermojunction, the other thermojunction of the electrical circuit being held just beneath the surface of water in which melting ice is floating. The appropriate voltage generated in the circuit is measured with a vernier potentiometer and the temperature can then be interpreted from standard tables. The maximum rate of electrical energy input to the heater is sufficiently small that the internal bulk gas temperature may be taken as the vessel wall temperature, and this is measured with a copper-constantan thermojunction. The walls of the vessel are water-cooled in order to stabilize conditions.

When the temperature difference between the surface of the heater element and the mild steel vessel is very small (<5 °C), a differential thermocouple between the two may need to be used in order to measure the temperature difference sufficiently accurately. The heat leaks through the thermoelements must then be estimated and appropriate corrections made if necessary.

Procedure. In a series of runs each with a different value of T_0, take measurements at various pressures of the power dissipation in the heater necessary to maintain a steady-state temperature difference, $(T - T_0)$, between the element and the water-cooled envelope. Draw the appropriate curves as in Figure 28.2.

In an alternative approach, obtain measurements for the heat leak *versus* steady-state temperature difference at constant pressure. (This avoids the almost continuous adjustment of power dissipation when the gas pressure is changed until a constant steady-state temperature difference can be maintained.) Repeat the procedure throughout the pressure range. By crossplotting, obtain the required family of $\dot{Q}$ *versus* p curves for a series of constant T values. This approach usually gives more consistent results.

Further Experiments

The same basic principles as are outlined above can be used for two other related projects:

(*a*) The variation of the heat leak between two parallel plates with their temperature difference, the separation and dimensions of the plates, the inclination of the plates to the vertical and air pressure.

(*b*) The rate of evaporation of liquid refrigerant (e.g. N_2(l) from a dewar vessel as a function of the vacuum interspace pressure [7].

Discussion

What can you infer from the measurements you have made? What sources of error are involved?

References

1. Carpenter, L. G. *Vacuum Technology*, Hilger, London, p. 96, 1970.
2. Maunder, L. and Probert, S. D. (editors) *Experimentation for Students of Engineering*, **1** Ch. 5, Appendices 5.1 and 5.2, Heinemann 1970.
3. Simonson, J. R. *Engineering Heat Transfer*, McGraw-Hill, Maidenhead, p. 214, 1967.
4. Kennard, E. H. *Kinetic Theory of Gases*, 1st edition, McGraw-Hill, New York, 1938.
5. Leck, J. H. *Pressure Measurement in Vacuum Systems*, 2nd edition, Chapman & Hall, London, pp. 40–49, 1964.
6. Probert, S. D., Brooks, R. G., Thomas, T. R. and Maxwell, J. 'Heat Losses through Rarefied Fluids', *Int. J. Heat Mass Transfer*, **10**, pp. 135–148, 1967.
7. Probert, S. D. *Thermal Insulation in Relation to Cryogenics*, H.M.S.O., p. 19, 1968.

Index